Tobias Wittkop

Transitivity Clustering

Tobias Wittkop

Transitivity Clustering

Clustering biological data by unraveling hidden transitive substructures

Südwestdeutscher Verlag für Hochschulschriften

Impressum/Imprint (nur für Deutschland/ only for Germany)
Bibliografische Information der Deutschen Nationalbibliothek: Die Deutsche Nationalbibliothek verzeichnet diese Publikation in der Deutschen Nationalbibliografie; detaillierte bibliografische Daten sind im Internet über http://dnb.d-nb.de abrufbar.

Alle in diesem Buch genannten Marken und Produktnamen unterliegen warenzeichen-, markenoder patentrechtlichem Schutz bzw. sind Warenzeichen oder eingetragene Warenzeichen der jeweiligen Inhaber. Die Wiedergabe von Marken, Produktnamen, Gebrauchsnamen, Handelsnamen, Warenbezeichnungen u.s.w. in diesem Werk berechtigt auch ohne besondere Kennzeichnung nicht zu der Annahme, dass solche Namen im Sinne der Warenzeichen- und Markenschutzgesetzgebung als frei zu betrachten wären und daher von jedermann benutzt werden dürften.

Verlag: Südwestdeutscher Verlag für Hochschulschriften Aktiengesellschaft & Co. KG
Dudweiler Landstr. 99, 66123 Saarbrücken, Deutschland
Telefon +49 681 37 20 271-1, Telefax +49 681 37 20 271-0
Email: info@svh-verlag.de
Zugl.: Bielefeld, Univeristy, Diss., 2010

Herstellung in Deutschland:
Schaltungsdienst Lange o.H.G., Berlin
Books on Demand GmbH, Norderstedt
Reha GmbH, Saarbrücken
Amazon Distribution GmbH, Leipzig
ISBN: 978-3-8381-1654-9

Imprint (only for USA, GB)
Bibliographic information published by the Deutsche Nationalbibliothek: The Deutsche Nationalbibliothek lists this publication in the Deutsche Nationalbibliografie; detailed bibliographic data are available in the Internet at http://dnb.d-nb.de.

Any brand names and product names mentioned in this book are subject to trademark, brand or patent protection and are trademarks or registered trademarks of their respective holders. The use of brand names, product names, common names, trade names, product descriptions etc. even without a particular marking in this works is in no way to be construed to mean that such names may be regarded as unrestricted in respect of trademark and brand protection legislation and could thus be used by anyone.

Publisher: Südwestdeutscher Verlag für Hochschulschriften Aktiengesellschaft & Co. KG
Dudweiler Landstr. 99, 66123 Saarbrücken, Germany
Phone +49 681 37 20 271-1, Fax +49 681 37 20 271-0
Email: info@svh-verlag.de

Printed in the U.S.A.
Printed in the U.K. by (see last page)
ISBN: 978-3-8381-1654-9

Copyright © 2010 by the author and Südwestdeutscher Verlag für Hochschulschriften Aktiengesellschaft & Co. KG and licensors
All rights reserved. Saarbrücken 2010

SUMMARY

Clustering is a computational technique for the assignment of objects into groups of similar elements. Generally, it is widely used for business data interpretation, natural language analyses, and image processing, just to name a few. Typical bioinformatic applications are: (1) detection of homologous proteins; single and multi domain, (2) prediction of protein complexes in protein-protein interaction networks, (3) identification of overrepresented DNA sequence patterns, and (4) gene co-expression studies.

Traditionally, we distinguish between partitional, overlapping, and hierarchical approaches. Partitional and overlapping approaches follow two different strategies: (1) center-based approaches for the detection of appropriate cluster representatives, such as k-means and (2) methods for the identification of homogeneous clusters, such as Markov Clustering. Hierarchical approaches allow for the construction of a tree structure; single linkage agglomerative clustering may serve as an example here.

Solving the following problems is crucial for a successful cluster analysis: (1) Probably most challenging is the identification of a problem-specific similarity function. (2) Every clustering approach incorporates at least one parameter that influences the size and number of the clusters. Determining such a density parameter strongly depends on the problem and the chosen similarity function. Preferably, one can even prove certain attributes of a clustering result, given a similarity function and the density parameter. (3) Currently, high throughput experiments produce huge amounts of data. Hence, a clustering environment has to be capable of processing hundreds of thousands of data objects. (4) The integration of existing knowledge into a cluster analysis is highly valuable for improving the clustering output. The integration of known assignments may serve as an example here. (5) It is clear that the method needs to be robust against noise and outliers. (6) From an end-user's point of view, integration with standard software, appropriate visualization capabilities, and easy-to-use evaluation methods are highly beneficial.

This thesis introduces Transitivity Clustering (TC) and its accompanying software framework TransClust, a method which addresses all of the aforementioned problems. It is a homogeneous partitioning method based on Weighted Transitive Graph Projection (WTGP), which aims for unraveling hidden transitive substructures in a given similarity graph deduced from a pairwise similarity measure. TC solves the aforementioned problems (2-5). The software implementation TransClust is an easy-to-use standalone and online application that solves the problems mentioned in (1,6). Furthermore, in TC, the density parameter can be chosen intuitively and the underlying weighted transitive graph projection model allows certain criteria of the clustering results to be proven. In addition, the model has been extended in order to allow for the following advanced features: (1) The integration of existing knowledge, for instance, by means of upper and lower bounds, (2) the computation of an hierarchical clustering, and (3) the calculation of overlapping clusterings. These extensions widen the applicability of TC and provide features that distinguish TC from other bioinformatics alternatives.

The flexibility of TC makes it suitable for various real-world applications. In this work, we concentrate on protein sequence clustering and the detection of protein complexes in protein-protein interaction networks, showing that TC outperforms the most-commonly used bioinformatics clustering techniques.

The software implementation of TC, TransClust, is available online at `http://transclust.cebitec.uni-bielefeld.de` as web application, as standalone tool, and as plugin for the standard network analysis tool Cytoscape. It provides results of similar or superior accuracy to those of alternative approaches. It is unique in that it features an easy-to-use clustering environment that contributes to all the important steps in a cluster analysis: (1) the choice and evaluation of a meaningful similarity function, (2) the detection of an appropriate density parameter, (3) the efficient computation of a clustering, and (4) the interpretation and evaluation of the clustering results.

CONTENTS

1. *Introduction* 7
   1.1 Motivation 7
   1.2 Applications 9
   1.3 Structure of the thesis 10
   1.4 Availability 11

2. *Background and related work* 13
   2.1 Definitions and notations 13
   2.2 Protein sequence similarity 14
       2.2.1 BLAST 14
       2.2.2 Similarity functions for sequence clustering 14
   2.3 Clustering 16
       2.3.1 Quality measures between two clusterings 16
   2.4 Clustering algorithms 19
       2.4.1 Hierarchical Clustering 20
       2.4.2 K-means 20
       2.4.3 Markov Clustering 21
       2.4.4 Spectral Clustering 22
       2.4.5 Restricted Neighborhood Search Clustering 23
       2.4.6 GeneRAGE 24
       2.4.7 Affinity Propagation 25
       2.4.8 Summary 26

3. *Transitivity Clustering* 29
   3.1 Transitive graph projection 30
   3.2 Data partitioning by using weighted transitive graph projection .. 32
   3.3 Extensions 34
       3.3.1 Upper and lower bounds 34
       3.3.2 Building hierarchies 36
       3.3.3 Identification of overlaps 40
   3.4 Algorithms solving the WTGPP 43
       3.4.1 Fixed parameter branch and bound strategy 43
       3.4.2 FORCE 43
       3.4.3 A greedy approach 44

	3.4.4	Integer Linear Programming	44
	3.4.5	Cluster Affinity Search Technique	45

4. **The Transitivity Clustering framework TransClust** 47
   4.1 Data import 48
   4.2 Clustering methods 48
       4.2.1 Layout-based heuristic 50
       4.2.2 Integration of the CAST algorithm 56
       4.2.3 Integration of the exact fixed parameter approach 56
       4.2.4 Post-processing 57
       4.2.5 Extensions and integration of existing knowledge 59
       4.2.6 Threshold determination and supporting analyses 59
   4.3 Availability 60
       4.3.1 Standalone application 60
       4.3.2 Cytoscape plugin 62
       4.3.3 Web application 64
   4.4 Evaluation of the integrated TransClust framework 64
       4.4.1 Data 65
       4.4.2 Optimizing the combination of methods in TransClust 66
       4.4.3 Influence of post-processing on accuracy 67
       4.4.4 Comparison against exact solution 67

5. **Evaluations of the Transitivity Clustering model** 71
   5.1 Single-domain protein sequence clustering 72
       5.1.1 Data 72
       5.1.2 Evaluation method 73
       5.1.3 Results 73
   5.2 Protein sequence clustering 74
       5.2.1 Data 76
       5.2.2 Comparison to different clustering methods 76
       5.2.3 Example threshold determination 77
       5.2.4 Integration of additional knowledge 79
   5.3 Clustering protein-protein interaction networks 79
       5.3.1 Data 81
       5.3.2 Robustness analysis 82
       5.3.3 Evaluation on experimental data 84
       5.3.4 Finding overlaps with Transitivity Clustering 87

6. **Integrated applications** 89
   6.1 MoRAine 89

	6.1.1	Transcription factor binding site annotation - A difficult and error-prone task 89
	6.1.2	Methods 90
	6.1.3	Results and discussion 91
6.2	CoryneRegNet 94
	6.2.1	Integration of Transitivity Clustering with CoryneRegNet 96
	6.2.2	Inter-species transfer of gene regulatory networks 96

7. Discussion 99
   7.1 Transitivity Clustering and TransClust 99
   7.2 Computational biology applications 101
   7.3 Integration in bioinformatics tools 102
   7.4 Future directions 102

8. Conclusion 105

Bibliography 106

Appendix   121

A. Publications & cooperations 125

B. TransClust data formats 127

C. MoRAine 1.0 131

D. Supplementary figures and tables 133

1. INTRODUCTION

1.1 Motivation

Clustering is one of the most widely used methods in computational biology [4]. It describes the assignments of objects into groups that share common traits. Thereby, a huge amount of data is combined in order to provide the data analyst with a compact overview of the data sets to be investigated. One popular example in the life sciences is the grouping of proteins into families and superfamilies. Usually, a similarity or distance measure between the objects of interest is necessary. Generally it is desired to construct groups, where objects of the same cluster show higher similarity to each other than to objects from different clusters.

There are three different kinds of clustering:

Partitional clustering: Each object is assigned to exactly one group.

Overlapping clustering: Objects can be assigned to multiple groups.

Hierarchical clustering: For each two clusters, either one is a complete subset of the other or they are disjoint. Hence, the results may be represented as a tree.

Most current approaches are limited to one of these three types. Hence, one has to decide in advance for the desired type of clustering. As a consequence, different clustering models are difficult to compare since they follow different strategies and optimize a different objective function. If it is not known in advance, which clustering type is best suited for the given problem, a flexible clustering model that can be used for any of the three types of clustering would be beneficial. It would allow for comparing the results of the three kinds instead of comparing completely different approaches.

All clustering approaches share common problems and requirements. Crucial parts of any cluster analysis are:

Similarity function: The similarity function is highly problem-specific and is probably most influential for the clustering result. The detection of an appropriate similarity function is hence the first step in each cluster analysis. Many clustering methods skip this step and assume a similarity measure is given.

Density parameter: Every partitional clustering method needs at least one parameter that influences the size and the number of the resulting clusters. The number of clusters K in the widely used K-means approach may serve as an example for such a density parameter. The choice of an optimal value for this parameter is challenging. It demands either a parameter that can be chosen intuitively or specialized methods for its detection. Additionally, a clustering model where attributes of the resulting clusters are provably connected to the density parameter is preferable, since it eases results interpretation.

Runtime and space efficiency: Clustering tasks, especially in computational biology, can become very complex. The detection of groups of similar protein sequences, for instance, means clustering of hundreds of thousands of sequences. In order to handle such huge amounts of data, a clustering methods has to be runtime and space efficient. It should further be capable of taking advantage of sparse data, as this might improve the running time enormously.

Robustness against outliers and noise: Similarities are often derived from experiments or heuristic approaches, and thus may not reflect reality perfectly. Single outliers, for instance due to miscalculated similarities, should not have a high influence on the clustering result. Furthermore, a clustering approach should still be able to produce meaningful results even for noisy data.

Clustering has been used in computer science for a long time. Consequently, various different approaches were developed over the years. Most of them are designed for a specific application. Hence, the corresponding similarity function and density parameter were optimized application specifically. Non-application-specific methods require techniques to evaluate a similarity function and to detect a reasonable density parameter. These difficulties are usually neglected and left to the end-user. The density parameter is particularly hard to specify and most methods help only little or not at all with this task.

Further highly valuable aspects, considered only partially by most approaches, are:

Interpretable results: A clustering result with unclear relation to the similarities of the objects is hard to interpret. One would have to understand the whole clustering process of the used method to get an impression about how the clusters emerged and what the cluster assignment actually means. Preferably, a clustering model guarantees certain attributes of the clustering output. These attributes are ideally coupled directly to the given similarity function and the density parameter, allowing the end-user to intuitively judge the changes in the output when altering the input parameters.

Integration of existing knowledge: Clustering is generally considered as unsupervised learning. However, it might be important to include additional information. If it is known that two objects have to be in one cluster or conversely, that two objects must not be assigned to the same group, the clustering model should consider this to avoid false assignments and to improve the clustering quality. Generally, a clustering method should be able to benefit from the experience and knowledge of its user with the specific application case.

Integration with standard software: Clustering is often only one step in a data analysis. All other parts require additional methods and software. For many applications, standard software exists. Moreover, some software environments were developed to be applicable for multiple tasks. The network analysis and visualization software Cytoscape [29] may serve as one example for such an environment. It is highly valuable for a clustering method, to be integrated in such standard software in order to avoid typical data pipelining and data integration problems (see e.g. [62] for further details); and may help to answer typical application-specific follow-up questions.

Visualization: If the size of the clustering problem is limited to a few objects, a graph-based visualization is a powerful tool. It aids with specifying a meaningful density parameter and with the identification of outliers. Moreover, the clustering results can be analyzed more efficiently and essential information can be extracted quickly.

Evaluation methods: Various steps in a cluster analysis influence the results' quality; choosing the similarity function and the density parameter are only two of them. It is necessary to evaluate each step in order to produce a good clustering for a specific application.

Reproducible results: Another crucial feature in data clustering is the reproducibility of the results. If a tool's output differs for multiple runs on the same data sets with the same set of parameters, it is hard to judge the result's quality.

Small number of user-defined parameters: Many algorithms are heuristics. Hence they need several parameters to control the trade-off between quality and speed. The number of essential parameters should be limited to avoid misuage.

All of the above criteria are important, but for a specific application it is essential that the results "make sense". This thesis will introduce Transitivity Clustering (TC) and its implementation TransClust. It will be described how TC contributes to all of the above mentioned issues. We will see, that TC can compete and even outperform commonly used approaches for specific applications in computational biology. TC is a flexible clustering model based on a graph modification problem, the Weighted Transitive Graph Projection Problem (WTGPP). It can be utilized for partitioning, overlapping, and hierarchical clustering. The clustering output has provable attributes that depend on a single intuitive density parameter. The TransClust framework is an efficient and easy-to-use implementation of TC, which is available as web application, as standalone program, and as plugin for the standard network analysis environment Cytoscape.

1.2 Applications

The value of efficient and accurate clustering methods is clarified in the following by means of some brief application examples in the life sciences.

Gene expression data: Mao et al. [56] recently used Markov Clustering (MCL) to cluster 16,293 genes of *Arabidopsis thaliana* based on their co-expression level. They identified 527 modules (clusters) of co-expressed genes.

Data reduction: In 2007, Cameron *et al.* [25] presented a strategy to reduce the size of sequence databases by storing only representative union-sequences for each cluster of similar sequences. As an example the sequence database GenBank [67] was reduced by 27%, resulting in a by 22% decreased search time, with no significant change in accuracy.

Protein complexes: Krogan *et al.* [52] investigated a protein-protein interaction network of the yeast *Saccharomyces cerevisiae*, consisting of 2,708 proteins and 7,123 interactions. By using MCL, 547 protein complexes with an average size of 4.9 were predicted.

Word sense disambiguation: Recently, Duan *et al.* [31] used a clustering approach based on a minimal spanning tree approximation to assign words to groups that have the same meaning. On a test set of 21 ambiguous keywords, their approach outperforms other commonly used unsupervised text mining methods by 2%.

Certainly, this list does not cover all possible applications for clustering. Various others exist in computational linguistic, economics, or medicine, just to name a few. The examples illustrate that clustering is an important research topic. Improvements are necessary to keep up with new developments and technologies, and hence with novel applications in large-scale, high-throughput data analysis.

1.3 Structure of the thesis

The remainder of this thesis is structured as follows: The next chapter concentrates on the clustering problem in general, gives some basic definitions, and presents all clustering methods used within the thesis. Subsequently, Chapter 3 introduces to the WTGPP including attributes of the problem and modifications to customize it for specific tasks. Existing algorithms for solving this graph modification problem are presented in this chapter as well. Following this, Chapter 4 describes the software framework TransClust. All implemented algorithms can be found here and the chapter guides through the different TransClust options. Since TransClust also incorporates algorithms developed by other scientists, the integration of these algorithms is also described here, as well as an evaluation of the performance of the heuristic methods. Chapter 5 describes experiments performed to evaluate the applicability of TC to typical biological problems. Aside from a comparison against existing methods, the modifications of the original model are also evaluated here. Further applications can be found in Chapter 6. TC was integrated with MoRAine and CoryneRegNet. MoRAine is a software that optimizes the information content of a Position Specific Scoring Matrix (PSSM) by using clustering algorithms. With the integrated TC, runtime and quality of MoRAine has been significantly improved. Chapter 7 and Chapter 8 complete this thesis with discussion and conclusion. Supplementary material can be found in the Appendix.

1.4 Availability

All components of the presented software framework are online available at http://transclust.cebitec.uni-bielefeld.de. The source code of TransClust, the three Cytoscape plugins, and example files are available for download on this page. A web application can be used to cluster small problem instances and an online tutorial guides through the usage of TransClust. The predecessor of TransClust, FORCE, can be found at http://www.cebitec.uni-bielefeld.de/comet/force, together with the data used in the evaluation study of Section 5.1. The application MoRAine (see Section 6.1) can be used as a web application or be downloaded from http://moraine.cebitec.uni-bielefeld.de. CoryneRegNet in can be found at http://www.coryneregnet.de. The network visualization and analysis software Cytoscape can be found at http://www.cytoscape.org. Most applications are implemented in Java and do not need any additional libraries. Java script must be activated in the browser for some features at the TransClust website.

2. BACKGROUND AND RELATED WORK

2.1 Definitions and notations

Throughout this work two kinds of graphs appear; an undirected simple graph and a directed graph:

Definition 2.1 (Undirected simple graph). An undirected simple graph $G = (V, E)$ consists of a set of nodes V and a set of edges $E \subseteq \binom{V}{2}$, where $\binom{V}{2}$ denotes the set of all two element subsets of V. Following this definition the edges are undirected and the graph contains no self-loops or multiple edges between two nodes.

Note that uv is used shortly for an unordered pair $\{u, v\} \in \binom{V}{2}$.

Definition 2.2 (Directed graph). A directed simple graph $G = (V, E)$ consists of a set of nodes V and a set of edges $E \subseteq V \times V$. Following this definition the edges are directed and the graph may contain self-loops but no multiple edges between two nodes.

Definition 2.3 (Induced subgraph). An induced subgraph $G' = (V', E')$ of a graph $G = (V, E)$ consists of a set of nodes $V' \subset V$ and a set of edges $E' \subset E$. In E' are exactly those edges that connect elements of V' and are present in E, i.e. $E' = E \cap \binom{V'}{2}$ for undirected graphs and $E' = E \cap (V' \times V')$ for directed graphs respectively.

In the following, definitions of connected components for these kinds of graphs are defined. As prerequisite, a path between two objects has to be defined first.

Definition 2.4 (Path). In a graph $G = (V, E)$ a path between two nodes $u, v \in V$ is a sequence of nodes $u = v_1, ..., v_n = v$ for which every connecting edge exists, i.e. for $1 \leq i \leq n-1$: $\{v_i, v_{i+1}\} \in E$ (for undirected simple graphs) and $(v_i, v_{i+1}) \in E$ respectively (for directed graphs).

Definition 2.5 (Connected component/strongly connected component). A connected component of an undirected simple graph $G = (V, E)$ is an induced subgraph $G' = (V', E')$ of G, where V' is the maximal subset of V, such that there exists a path between every two nodes in G'. For a directed graph the definition is the same but G' is called strongly connected.

Definition 2.6 (Weakly connected component). A weakly connected component of a directed graph $G = (V, E)$ is an induced subgraph $G' = (V', E')$ of G, which consists of a strongly connected component $G'' = (V'', E'')$ and all nodes $u \in V$ that are connected to G'', i.e. there exists a path between u and one node $v \in V''$ in G'.

The last necessary definition used throughout this thesis is the transitivity of an undirected graph.

Definition 2.7 (Transitivity). An undirected simple graph $G = (V, E)$ is called transitive if for all triples $uvw \in \binom{V}{3}$, $uv \in E$ and $vw \in E$ implies $uw \in E$.

2.2 Protein sequence similarity

An application that repeatedly occurs in this work is the clustering of protein sequences. This section will introduce to BLAST [2], a heuristic for local alignments of DNA and amino acid sequences. The results from BLAST can be used to define a symmetric pairwise similarity function between protein sequences. Five such methods have been developed and will be described here.

2.2.1 BLAST

The Basic Local Alignment Search Tool (BLAST), developed by Altschul et al. [2] is a commonly used tool to compare sequences. BLAST uses a heuristic to find local alignments between a query sequence and a database of subject sequences. First, small sequences are located (default length 11) and afterwards extended to a local alignment if beneficial. It can also be used to compare protein sequences against each other in an all vs. all analysis. For each local alignment between subsequences of two proteins, an High Scoring Pair (HSP) is reported. Note that the alignment may differ for the two possible directions, due to BLAST's heuristic nature. Further note, that multiple HSPs may occur for the same pair of proteins for the same direction. In the 12 column tabular output (-m 8 option), BLAST lists HSPs as follows: (1) the name of the query protein, (2) the name of the subject protein, (3) the percent identity, (4) the alignment length, (5) the number of mismatches, (6) the number of gaps, (7) the start of the query sequence, (8) the end of the query sequence, (9) the start of the subject sequence, (10) the end of the subject sequence, (11) the E-value; the number of hits that are expected to occur by chance in a database of the same size with at least the reported Bit-score, i.e. a value for the significance of the alignment, and (12) the Bit-score of the alignment. To limit running time one can specify an E-value threshold.

2.2.2 Similarity functions for sequence clustering

Assume it is given a set of proteins V and a BLAST output file possibly containing multiple HSPs in both directions. For two proteins u and v let $(u \leftarrow v)_i$ and $(u \rightarrow v)_j$, where $i = 1, \ldots, k$ and $j = 1, \ldots, l$ be the corresponding k HSPs in one and l HSPs in the other direction, respectively. If no HSPs exist in at least one of the two directions, any of the following similarities is set to the minimal value 0.

2.2. PROTEIN SEQUENCE SIMILARITY

Best Hit (BeH) This widely used method concentrates on the E-value of a single HSP: For both directions, one looks for the best hit, i.e., the HSP with lowest E-value. To obtain a symmetric similarity function sim: $\binom{V}{2} \to \mathbb{R}$, the negative logarithm of the worst (largest) of the two E-values is taken as similarity measure between two sequences u and v. The resulting symmetric similarity function is then defined as:

$$\text{sim}(uv) := -\log_{10}\left(\max\left\{\min_{i=1,\ldots,k} \text{E-value}\left((u \leftarrow v)_i\right), \min_{j=1,\ldots,l} \text{E-value}\left((u \to v)_j\right)\right\}\right).$$

Sum of Hits (SoH) This approach is similar to BeH, but additionally includes every HSP between two sequences. This may be useful if multiple unconnected subsequences are important for the protein assignment; one example is the problem of protein domain shuffling [71]. The SoH similarity function is defined as:

$$\text{sim}(uv) := -\log_{10}\left(\max\left\{\prod_{i=1}^{k} \text{E-value}\left((u \leftarrow v)_i\right), \prod_{j=1}^{l} \text{E-value}\left((u \to v)_j\right)\right\}\right).$$

Coverage (Cov) The third approach integrates the lengths and the sequence identity of an HSP into the similarity function. To determine the coverage, the following indicator function is needed:

$$\mathbb{I}_{uv}(i) := \begin{cases} 1 & \text{if in } u \text{ the position } i \text{ is covered by any HSP } (u \leftarrow v)_{1 \leq n \leq k} \\ & \text{or } (u \to v)_{1 \leq m \leq l} \text{ respectively} \\ 0 & \text{otherwise.} \end{cases}$$

The coverage can now be defined as

$$\text{coverage}(uv) := \min\left(\frac{1}{|u|}\sum_{i=1}^{|u|} \mathbb{I}_{uv}(i), \frac{1}{|v|}\sum_{i=1}^{|v|} \mathbb{I}_{vu}(i)\right).$$

In order to obtain a good similarity function, the influence of the coverage on the overall similarity function is controlled by a factor f, and set to:

$$\text{sim}(uv) := \text{sim}'(uv) + f \cdot \text{coverage}(uv).$$

where $\text{sim}' \colon \binom{V}{2} \to \mathbb{R}$ denotes one of the previously presented similarity functions, BeH or SoH.

Score BLAST calculates a bit score for each HSP depending on the number of necessary insertions and deletions for the alignment. A common approach to define a pairwise similarity between two sequences is to use this score and normalize it to the length of the HSP. It is recommended to filter the list of HSPs for hits with reliable E-values. Otherwise the normalization may lead to high similarities of dissimilar objects, if a common subsequence is very short. As in the previously introduced measure BeH, the maximal score for one direction is used if multiple HSPs exist. Symmetry of the similarity function is achieved by choosing the lower score of both directions. The similarity based on the normalized score is defined as follows:

$$\text{sim}(uv) := \max \left\{ \min_{i=1,\ldots,k} \frac{\text{Score}\left((u \leftarrow v)_i\right)}{\left|(u \leftarrow v)_i\right|}, \min_{j=1,\ldots,l} \frac{\text{Score}\left((u \rightarrow v)_j\right)}{\left|(u \rightarrow v)_j\right|} \right\}$$

2.3 Clustering

Clustering is a common computational technique for data analysis in the life sciences. Essentially, it is the assignment of objects into groups, where objects within a group are more similar to each other than objects between two groups. One can distinguish between three types of clusterings: (1) Partitional clustering divides the set of objects into disjoint groups, (2) overlapping clustering allows assignments of objects to multiple groups, and (3) hierarchical clustering defines hierarchical sets, where two clusters are either disjoint, or one is a complete subset of the other. The formal definitions are:

Partitional clustering: A *partitional clustering* of a set of objects S is a subset $S' = \{s_1, \ldots, s_n\}$ of the power set $\mathcal{P}(S)$, such that $S = \bigcup_{i=1}^n s_i$ and $s_i \cap s_j = \emptyset$, $1 \leq i \neq j \leq n$.

Overlapping clustering: An *overlapping clustering* of a set of objects S is a subset $S' = \{s_1, \ldots, s_n\}$ of $\mathcal{P}(S)$, such that $S = \bigcup_{i=1}^n s_i$

Hierarchical clustering: An *hierarchical clustering* of a set of objects S is a subset $S' = \{s_1, \ldots, s_n\}$ of $\mathcal{P}(S)$, such that $S = \bigcup_{i=1}^n s_i$ and for each pair of clusters s_i, s_j $1 \leq i \neq j \leq n$ one of the following conditions holds:

- $s_i \cap s_j = \emptyset$
- $s_i \subset s_j$
- $s_j \subset s_i$

2.3.1 Quality measures between two clusterings

One method for evaluating the quality of a clustering is to compare it against a gold standard assignment. This external quality evaluation allows to compare different approaches, which optimize different internal optimization functions. For this purpose, various methods have been developed, like calculating the Positive Predictive Value (PPV) or the sensitivity between a clustering and a gold standard reference.

Some of the evaluations performed in Chapter 5 are based on previous studies. Consequently, the quality measures are the same as in the corresponding evaluations. This section introduces all measures used within this work.

2.3. CLUSTERING

In the following let $C = \{C_1, \ldots, C_n\}$ be the clustering obtained from the algorithm and $K = \{K_1, \ldots, K_m\}$ be the reference clustering. Furthermore let $T = (t_{i,j}) \in \mathbb{N}^{m \times n}$ denote the matrix where each entry is the number of common objects between K_i and C_j.

$$t_{i,j} = |\{K_i \cap C_j\}| \; ; 1 \leq i \leq m, 1 \leq j \leq n$$

This section uses also the standard abbreviations for True Positives (TP), True Negatives (TN), False Positives (FP), and False Negatives (FN). Another notation is $|T|$ for the sum of all entries in T and $|T_{\cdot j}|$ and $|T_{i \cdot}|$ for the sums of the entries of the i-th row and j-th column respectively.

According to the evaluation of clustering algorithms for Protein-Protein Interaction (PPI) networks in [21], the following definitions are given:

Definition 2.8 (Positive predictive value). The PPV is generally defined as:

$$\text{PPV} = \frac{\text{TP}}{\text{TP} + \text{FP}}$$

This value ranges from zero to one, and describes the ratio of correct predictions to all predictions. It can only reach the highest value of 1, if no false positive predictions occur, i.e. in the case of clustering: no pair of objects that belong to different clusters in the gold standard are assigned to the same cluster. Since the reference clustering may be overlapping, the PPV for each pair of clusters C_j, K_i is defined as:

$$\text{PPV}(C_j, K_i) = \frac{t_{i,j}}{|T_{\cdot j}|}$$

A cluster-wise PPV can then be defined for each cluster C_j as:

$$\text{PPV}(C_j) = \max_{1 \leq i \leq m} \text{PPV}(C_j, K_i)$$

To get an overall PPV between two clusterings, the cluster-wise PPVs are incorporated as follows:

$$\text{PPV}(C, K) = \frac{\sum_{j=1}^{n} \text{PPV}(C_j) \cdot |T_{\cdot j}|}{|T|}$$

Definition 2.9 (Sensitivity). Sensitivity is a value that reflects the quantity of the correct predictions. In the case of clustering this is the ratio between objects that are in the same cluster, both in the reference clustering and in the obtained clustering, against all objects in the reference clustering. Generally it is defined as:

$$\text{Sen} = \frac{TP}{TP + FN}$$

First a reference-cluster-wise sensitivity is defined as follows:

$$\text{Sen}(K_i) = \max_{1 \leq j \leq n} \frac{t_{i,j}}{|K_i|}$$

A general sensitivity can then be calculated by:

$$\text{Sen}(C, K) = \frac{\sum_{i=1}^{m} |K_i| \cdot \text{Sen}(K_i)}{|K|}$$

Definition 2.10 (Accuracy). The accuracy is a trade-off between PPV and sensitivity. Both values on their own can be high even if the clustering is not perfect. A clustering with only singletons would lead to a high PPV since no false positive prediction occur, while building one big cluster containing all elements would have the maximal sensitivity value. Neither of these examples is necessarily desirable. Hence, a combination of these values, the accuracy, evaluates the quality better.

The arithmetic Accuracy (ACC) is the arithmetic mean of sensitivity and PPV.

$$\text{ACC}_{arit}(C, K) = \frac{\text{Sen}(C, K) + \text{PPV}(C, K)}{2}$$

The geometric ACC takes the geometric mean of sensitivity and PPV.

$$\text{ACC}_{geo} = \sqrt{\text{Sen}(C, K) \cdot \text{PPV}(C, K)}$$

A novel distance measure has been introduced by Brohée et al. [21] and is called separation. In contrast to the above described values it takes all pairwise relations between the clusters obtained from the algorithm and the reference clusters into account, and does not concentrate on the best matching cluster.

Definition 2.11 (Separation). The separation for each pair of clusters C_j and K_i is defined as:

$$\text{Sep}(C_j, K_i) = \frac{t_{i,j}^2}{|T_{i,\cdot}| \cdot |T_{\cdot,j}|}$$

Now it is possible to define the separation for each cluster C_j and K_i as the sum of pairwise separations:

$$\text{Sep}(C_j) = \sum_{i=1}^{m} \text{Sep}(C_j, K_i)$$

and

$$\text{Sep}(K_i) = \sum_{j=1}^{n} \text{Sep}(C_j, K_i)$$

To obtain an overall value for the two clusterings C and K one takes the mean of separations for all clusters in C (cluster-wise) and K (reference-cluster-wise) and subsequently calculates

2.4. CLUSTERING ALGORITHMS

the geometric mean.

$$\begin{aligned}\operatorname{Sep}(C,K) &= \sqrt{\frac{\sum_{j=1}^{n} \operatorname{Sep}(C_j)}{|C|} \cdot \frac{\sum_{i=1}^{m} \operatorname{Sep}(K_i)}{|K|}} \\ &= \frac{\sum_{j=1}^{n} \sum_{i=1}^{m} \operatorname{Sep}(C_j, K_i)}{\sqrt{m \cdot n}} \\ &= \frac{\sum_{j=1}^{n} \sum_{i=1}^{m} \frac{t_{i,j}^2}{|T_{i,\cdot}| \cdot |T_{\cdot,j}|}}{\sqrt{m \cdot n}}\end{aligned}$$

Definition 2.12 (F-measure). The last quality measure is based on the general definition of Recall, Precision, and the F-measure:

Recall: $\frac{TP}{TP+FN}$

Precision: $\frac{TP}{TP+FP}$

F-measure: $2 \cdot \frac{\text{Precision} \cdot \text{Recall}}{\text{Precision}+\text{Recall}} = \frac{2 \cdot TP}{(TP+FP)+(TP+FN)}$

Paccanaro et al. [61] modified the F-measure for comparing a clustering C against a reference clustering K in the following way:

First the best cluster C_j for each cluster K_i of the reference is found with respect to the standard definition of F-measure:

$$\text{F-measure}(K_i) = \max_{1 \le j \le n} \frac{2 \cdot t_{i,j}}{|C_j| + |K_i|}$$

The overall F-measure is then defined as:

$$\begin{aligned}\text{F-measure}(C,K) &= \frac{1}{\sum_{i=1}^{m} |K_i|} \sum_{i=1}^{m} \left(|K_i| \cdot \text{F-measure}(K_i)\right) \\ &= \frac{1}{\sum_{i=1}^{m} |K_i|} \sum_{i=1}^{m} \left(|K_i| \cdot \max_{1 \le j \le n} \frac{2 \cdot t_{i,j}}{|C_j| + |K_i|}\right)\end{aligned}$$

2.4 Clustering algorithms

Many clustering algorithms exist based on different models. While some try to find a list of center nodes for a partitional clustering, others aim to identify homogeneous clusters. Some partition the data and others create an hierarchical or overlapping clustering. For this task most algorithms try to optimize an objective function, that reflects the actual clustering aim. Since these optimizations often have a high problem complexity, heuristic methods are frequently used. As a consequence, most methods require several input parameters, which determine the quality of the used heuristic. Since clustering is an unsupervised data mining technique, all approaches have to have at least one parameter: the density parameter, which influences the number and size of the resulting clusters. This parameter can be as specific as the exact number of expected clusters, or more abstract like the level in a hierarchy from which a partition is created. This section will introduce the most-commonly used clustering techniques in computational biology. For each approach, the underlying model is presented together with the corresponding objective, a description of the density parameter, and the used algorithm.

2.4.1 Hierarchical Clustering

There are mainly two different kinds of hierarchical clustering; bottom-up approaches, where a set is iteratively divided into subsets, and top-down approaches, which start with only singletons and merge them iteratively. Subsequently, the focus lies on the second kind, which is commonly referred to as agglomerative clustering.

Given a set of objects V and a pairwise similarity sim: $\binom{V}{2} \to \mathbb{R}$, the main idea is to start with singletons as initial clustering and merge those two clusters which are the most similar. While in the first step the similarity is given by a pairwise similarity between the objects, in later iterations it is necessary to define a similarity between two clusters. Given a partitional clustering $C = \{C_1, \ldots, C_k\}$, which initially contains clusters of size 1, the three most popular similarity functions between two clusters C_i and C_j are:

Single linkage: $\mathrm{s}(C_i C_j) := \max\{\mathrm{sim}(uv); u \in C_i, v \in C_j\}$

Complete linkage: $\mathrm{s}(C_i C_j) := \min\{\mathrm{sim}(uv); u \in C_i, v \in C_j\}$

Average linkage: $\mathrm{s}(C_i C_j) := \frac{1}{|C_i| \cdot |C_j|} \sum_{u \in C_i} \sum_{v \in C_j} \mathrm{sim}(uv)$

To get a partitional clustering, given an hierarchical clustering, one can either select a set of cut nodes, which each represent a cluster, or specify a level in the hierarchy. One drawback of hierarchical clustering is that once an object is assigned to a cluster this decision is final. One popular example of hierarchical clustering in computational biology is the SYSTERS [51, 57] database, where single linkage clustering is one step in the assignment of proteins into families and superfamilies.

2.4.2 K-means

K-means is a standard data partitioning method. For a given set of objects $V = \{v_1, \ldots, v_n\}$ and a position $p_i \in \mathbb{R}^l$ ($l \geq 1$) for each object, the problem is defined as follows:

Find that partitioning $C = \{C_1, \ldots, C_K\}$ of V, such that the sum of squares of distances between the objects in one cluster and the cluster mean m_j is minimized:

$$C = \underset{C'}{\operatorname{argmin}} \left(\sum_{i=1}^{K} \sum_{v_j \in C'_i} \|p_j - m_i\|^2 \right)$$

The problem is known to be NP-hard[1] in an Euclidian space [1]. Subsequently described is

---
[1] A decision problem is in NP if it can be solved in polynomial time with a non-deterministic Turing Machine. P are the decision problems that can be solved in polynomial time with a deterministic Turing Machine. A problem p is NP-hard if every other problem in NP can be reduced in polynomial time to p. Unless P=NP, no NP-hard problem can be solved in polynomial time with a deterministic Turing Machine.

2.4. CLUSTERING ALGORITHMS

the commonly used Lloyd's algorithm, a heuristic to solve this problem [55].

Starting with an initial selection of K seed positions, all objects are assigned to their closest seed. Different methods can be used to define an initial seed selection, for instance, a random choice within the range of the objects' positions. After the first clustering has been calculated, the mean of each cluster is used as a seed for the next iteration. If only distances between the objects are used, that seed which has the smallest distance to all other elements of its cluster is chosen as a new seed. These steps of reassignment of the objects to clusters and calculating new seed positions is repeated until a stable stage is reached, i.e. no changes in the clustering appear anymore. Depending on the choice of initial seeds the result may vary. Consequently, having random initial seed nodes may lead to different results for every run. A possibility to overcome or reduce this problem is to either allow multiple runs with different initial seed positions, or to develop a method to "guess" a good choice of initial center nodes.

Note that K-means can also be applied if only a distance or similarity between the objects is defined. In this case the cluster mean (seed) is one of the nodes in the respective cluster.

2.4.3 Markov Clustering

The Markov Clustering (MCL) algorithm, developed by Stijn van Dongen [76], finds cluster structures in graphs by utilizing a mathematical bootstrapping procedure. The algorithm works on a column stochastic matrix:

$$M = (m_{i,j}) \in [0,1]^{n \times n}, \text{ where } \sum_{j=1}^{n} m_{i,j} = 1 \text{ for all } 1 \leq i \leq n$$

Given a pairwise similarity function, the entries of M are calculated by adding a loop to each node, i.e. define a self-similarity if not present, and normalize each column such that it is stochastic. This matrix is transformed by alternating two operations until an equilibrium state is reached. The first operation is called expansion and simulates a random walk on M as a simple matrix multiplication:

$$M \rightarrow M^2$$

Inflation, the second operation, transforms the columns individually. Each entry in a column is taken to the power of a parameter k, the inflation parameter. Afterwards the columns are normalized, such that the matrix is again column stochastic.

$$m_{i,j} \rightarrow \frac{m_{i,j}^k}{\sum_{i=1}^{n} m_{i,j}^k}$$

The inflation increases the difference in affinity between the objects, i.e. higher similarities become bigger while small similarities become smaller. This effect can be controlled by the inflation parameter, where larger values increase the difference between the similarities more than smaller values. The choice of the inflation parameter has a high impact on the clustering results, i.e. the granularity, and serves consequently as the density parameter for this approach.

In the equilibrium state the clusters are defined as the weakly connected components of the directed graph $G = (V, E)$ induced by the final matrix. In G a directed edge between two nodes i and j exists if $m_{i,j} > 0$. The so defined clusters can overlap, though in practice these overlaps rarely occur and are eliminated by merging the two corresponding clusters.

MCL is a fast method with only one crucial parameter. A detailed description about the convergence of this algorithm can be found in [76]. To further improve the runtime, the implementation by van Dongen takes only the best n neighbors of a node into account, where n is a variable parameter. The so derived sparse matrices can be computed much faster at the expense of accuracy.

MCL has been widely used for the identification of homologous proteins[2] (TribeMCL [32]) as well as for the detection of orthologous proteins (OrthoMCL [54]). Since it performed best in various comparisons between different clustering approaches, for instance [21, 78], it is an ideal candidate for evaluating against other clustering methods.

2.4.4 Spectral Clustering

In Spectral Clustering (SC) algorithms, the clusters are calculated by transforming the initial similarity matrix into a matrix of eigenvalues and subsequently applying K-means on this matrix. Consequently the number of expected clusters K has to be specified in advance. SC works as follows:

A given similarity matrix $S = (s_{i,j}) \in \mathbb{R}^{n \times n}$ is first normalized:

$$S' = D^{-\frac{1}{2}} S D^{-\frac{1}{2}}$$

where $D = (d_{i,j}) \in \mathbb{R}^{n \times n}$ is a diagonal matrix defined as:

$$D = diag(d_1, \ldots, d_n); \ d_i = \sum_{j=1}^{n} s_{i,j}$$

Afterwards one computes the matrix of eigenvalues $U = (u_{i,j}) \in \mathbb{R}^{n \times k}$, where the column u_i corresponds to the K largest eigenvalues of S'. Each row of U will be normalized again and one ends up with a matrix:

$$Y = (y_{i,j}) \in \mathbb{R}^{n \times K}, \text{ with } y_{i,j} = \frac{u_{i,j}}{\sqrt{\sum_{j=1}^{n} u_{i,j}^2}}$$

Finally each row of Y is treated as a point in $\mathbb{R}^K$ and will be clustered using a K-means algorithm.

The version of SC that is used for the evaluation in section 5.2 is a MATLAB implementation by Paccanaro et al. [61]. It has been developed for the task of protein sequence clustering. According to their publication the best number of clusters K has been determined by calculating

---

[2] Two genes/proteins are homologous if they are evolutionary related. One distinguishes between orthologous genes that occur due to a speciation event and paralogous genes that were created by gene duplication.

2.4. CLUSTERING ALGORITHMS

the eigenvalues λ_i of $M = S \cdot D^{-1}$, where $\lambda_i > \lambda_{i+1}$ for each $1 \leq i \leq n$ by:

$$K = min\{i; \frac{\lambda_i}{\lambda_{i+1}} > \epsilon\}$$

where ϵ is a predefined threshold.

SC has similar advantages and disadvantages as K-means due to its density parameter and its objective function. Since it performed best in a recent comparison between different protein domain sequence clustering approaches [61], besides MCL it is a further candidate for subsequent evaluations.

2.4.5 Restricted Neighborhood Search Clustering

In 2004, King et al. [49] developed a cost-based clustering approach for the task of predicting complexes of proteins in a PPI network. The Restricted Neighborhood Search Clustering (RNSC) approach uses only non-weighted similarities, i.e. a similarity function with values in $\{0, 1\}$ that decides whether two elements are similar or not. For a set of elements V the undirected graph $G = (V, E)$ is naturally defined using these similarities. Generally the goal is to find the clustering C that minimizes the cost function

$$\text{cost}(G, C) = \frac{n-1}{3} \sum_{v \in V} \frac{(\text{inc}(G, C, v) + \text{nadj}(G, C, v))}{|N(v) \cup C_v|}$$

where $\text{inc}(G, C, v)$ denotes the number of cross-edges incident to v, and $\text{nadj}(G, C, v)$ are the number of vertices in C that are in the same cluster C_v as v but not adjacent to it. $N(v)$ is the open neighborhood of v. For a faster approximation a second cost function is defined as:

$$\text{naive-cost}(G, C) = \frac{1}{2} \sum_{v \in V} (\text{inc}(G, C, v) + \text{nadj}(G, C, v))$$

Starting with an initial clustering, which can be either random or derived by other clustering methods, the RNSC algorithm first tries to optimize the naive-cost function by moving nodes from one cluster to another. After reaching an optimum for the naive-costs, RNSC again moves nodes from one cluster to another, but this time tries to optimize the actual cost function. The moves made by RNSC are either moves to decrease the costs, or random moves to avoid ending in a local minimum. Furthermore, the algorithm forbids processing a node after it has been moved until a certain number of other nodes has been moved. This avoids circling around a local minimum. For every move it is allowed to empty a cluster or create a new singleton cluster as long as the defined maximal number of clusters is not exceeded. Due to the random moves and the random initial clustering the results can differ for different runs of the algorithm.

Many parameters have to be specified in this method, namely: (1) the diversification frequency for how often diversification movements should be applied to avoid ending up in a local minimum; (2) the shuffling diversification length, which is the number of random diversification movements that are performed in each iteration; (3) the tabu length, which defines the size of the tabu list, that specifies if a node is allowed to be moved again; (4) the tabu list tolerance, which is a value for how often an object can appear in the tabu list before it is set to 'not movable'; (5) the number of experiments defining how often the clustering should be repeated, since the random aspect leads to different results for every run; (6) the naive stopping tolerance, which defines the number of moves that are allowed to be made without making any improvements to the naive cost function; (7) the scaled stopping tolerance, which is similar to the naive stopping tolerance but used for the accurate cost function and (8) the maximal number of clusters, which is simply an upper bound for the clusters that can be produced. Although many parameters effect the results quality, apparently the highest impact has the user-specified maximal number of clusters; the density parameter.

2.4.6 GeneRAGE

Enright et al. presented GeneRAGE, a protein sequence clustering algorithm [33]. GeneRAGE also searches for multi-domain proteins that may belong to several families thus producing an overlapping clustering. The main steps in the algorithm are:

Create directed similarity graph $G = (V, E)$: First, the CAST algorithm developed by Promponas et al. [63] is applied as a filtering algorithm (not to be confused with the Cluster Affinity Search Technique (CAST) approach) on BLAST results to remove biased hits. A directed similarity graph $G = (V, E)$ is built afterwards using a threshold t. Only the similarities above the threshold are taken into account, leading to an edge set $E \subseteq (V \times V)$ of:
$$E = \{(u, v) \in V \times V; \mathrm{sim}(u, v) > t\}.$$
Note that the similarity function at this point is not required to be symmetric.

Add/remove non-symmetric edges: For every two nodes u, v of V it is checked if both directions are present in the edge set; $(u, v) \in E$ and $(v, u) \in E$. If one direction exceeds the threshold but the other does not a Smith-Waterman alignment [70] between the sequences is performed again. The missing edge is added if the Z-Score of the alignment is above 10 and the present edge is removed otherwise. The resulting graph is symmetric and can consequently be written as an undirected graph $G' = (V, E')$.

Search for multi-domain proteins: For every three proteins $uvw \in \binom{V}{3}$ with $uv, uw \in E'$ and $vw \notin E'$ a second check with a Smith-Waterman alignment between the sequences v and w is performed. This check should indicate if the missing edge is a false negative. If the Z-score for this alignment is below 10, u is added to the list of putative multi-domain proteins, since it is connected to two groups, that of v and that of w.

Clustering: Single linkage clustering is applied on the undirected graph G', i.e. the connected components of G' are reported as clusters.

2.4. CLUSTERING ALGORITHMS

Split clusters with multi-domain proteins: Clusters obtained from the previous step are separated into smaller clusters, if they contain multi-domain proteins that act as connection between two families. These connecting proteins are assigned to both clusters, leading to an overlapping clustering.

GeneRAGE has been designed for clustering protein sequences. The density parameter of this approach is the similarity threshold used to create the initial directed graph G, though other parameters might have an influence as well, such as the Z-score threshold to validate an edge.

2.4.7 Affinity Propagation

Another centralized clustering algorithm is Affinity Propagation (AP) [34]. By passing messages between the data points, centers of clusters are determined and each data point is assigned to its closest center. Hereby AP tries to maximize the net-similarity S, defined as:

$$S = \sum_{i=1}^{n} \text{sim}(i, c_i) + \sum_{i=1}^{n} \delta_i$$

where c_i is the center corresponding to node i and δ_i is $-\infty$ if a node is not its own center ($i \neq c_i$), but there exist other nodes with center i ($\exists j,\ 1 \leq j \leq n;\ i = c_j$) and 0 otherwise. To optimize this objective function two kinds of messages are passed between the data points:

Responsibility $r(i, k)$ is sent from data point i to data point k reflecting how well k would act as center/exemplar for i with respect to all other possible centers of i.

Availability $a(i, k)$ is sent from data point k to data point i, giving evidence of how well k is suited to be a center of i using information about other nodes which may have k as center.

The process starts by initializing all availabilities to zero. The self-responsibilities $r(k, k)$ are the density parameter of this clustering approach. The values for responsibility are determined by:

$$r(i, k) = \text{sim}(i, k) - \max_{1 \leq k' \leq n; k' \neq k} \{a(i, k') + \text{sim}(i, k')\}$$

The availabilities for $i \neq k$ are calculated by:

$$a(i, k) = \min\left\{0, r(k, k) + \sum_{1 \leq i' \leq n; i' \notin \{i, k\}} \max\{0, r(i', k)\}\right\}$$

and the self-availabilities are calculated by:

$$a(k, k) = \sum_{1 \leq i' < n; i' \notin \{k\}} \max\{0, r(i', k)\}$$

It is further necessary to specify an additional parameter, the damping factor $\lambda \in [0, 1]$. To avoid numerical oscillations, the change of the values of responsibility and availability is limited with this factor by:

$$r(i, k) = r(i, k)_{\text{current}} \lambda + r(i, k)_{\text{previous}} (1 - \lambda)$$

where $r(i,k)_{\text{current}}$ is the responsibility for the current iteration, calculated as described above and $r(i,k)_{\text{previous}}$ is the responsibility of the previous iteration. The same applies to the availability:
$$a(i,k) = a(i,k)_{\text{current}} \lambda + a(i,k)_{\text{previous}}(1-\lambda)$$
During an iteration of AP the following steps are performed:

- Calculate all responsibilities.

- Calculate all availabilities.

- Assign all nodes i to that center k that maximizes $a(i,k) + r(i,k)$:

$$c_i = \text{argmax}_{1 \leq k \leq n} a(i,k) + r(i,k)$$

The algorithm stops either if a user-defined maximal number of iteration is reached, if changes of responsibilities and availabilities between the iterations fall below a user defined threshold, or if the obtained center assignments do not change over a certain number of iterations. Due to the multiple stopping criteria a convergence of the algorithm is not guaranteed. According to the developers AP is capable of clustering 20,000 objects in less than a day (120,000 if the similarity matrix is very sparse and the calculations of the message values are restricted to points where a similarity is defined).

2.4.8 Summary

Throughout this section it became obvious that one of the main challenges for each clustering approach is to define a meaningful density parameter to define the number/size of clusters. Given an hierarchical clustering, a set of cut nodes or the level in the hierarchy are such parameters to derive a partitioning. As there are various clustering approaches, there exist many different kinds of density parameters. Some approaches are limited to one value that has to be specified while others use a set of parameters. Using these values is not equally intuitive. While the number of expected clusters as in SC or K-means are intuitive values, the self-responsibility in AP or the inflation factor in MCL are more abstract. Often more than one value has to be specified. Heuristic approaches have additional parameters that influence the quality of the results according to their objective function. Table 2.1 summarizes the different values for the presented algorithms and specifies whether the algorithms are heuristic.

Table 2.2 illustrates how the presented approaches fulfill the desired features (as described in the Introduction).

2.4. CLUSTERING ALGORITHMS

	Density parameter	Further parameters	Strategy	Heuristic	Implementation	Availability
Hierarchical Clustering	set of cut nodes or hierarchy level	choice of similarity between clusters and elements (min,max,average)	assign nodes to closest cluster and thus create hierarchy	NO	libraries for many programming languages available	online, free
K-means	number of clusters K	possible: fixed start positions of center nodes	minimize sum of squares of distances to center nodes	YES	libraries for many programming languages available	online, free
Markov Clustering	inflation parameter	maximal number of neighbors, lower bound	alternate matrix multiplication and inflation until stable state	both	C++ an libraries for many programming languages	online, free
Spectral Clustering	number of clusters	different definitions of Laplace matrix	K-means on a matrix of Eigenvectors	YES	MATLAB	upon request
Restricted Neighborhood Search Clustering	Maximal number of clusters	7 different parameters	optimize cost function	YES	C++	upon request
GeneRAGE	similarity threshold	Z-score threshold of Smith-Waterman alignment	detection of protein families and multi-domain proteins	NO	C++	upon request
Affinity Propagation	self responsibility	damping factor, iterations	maximize net-similarity	YES	MATLAB and C++, web application	online, free

Tab. 2.1: Comparison of different clustering approaches.

Tab. 2.2: Overview of a variety of clustering algorithms and how they fulfill the desired features as specified in the requirement analysis in Chapter 1. *: includes method to evaluate or construct a similarity function, +: intuitive density parameter or methods included to detect a meaningful density parameter, !: provable attributes in the output given the similarity function and a density parameter. Abbreviations: HC: Hierarchical Clustering, MCL: Markov Clustering, SC: Spectral Clustering, RNSC: Restricted Neighborhood Search Clustering, AP: Affinity Propagation

Feature	HC	K-means	MCL	SC	RNSC	GeneRAGE	AP
Similarity function*		✓	✓	✓		✓	✓
Density parameter+							
Runtime and space efficiency		(✓)	✓	✓	✓	✓	✓
Robustness against outliers and noise			✓		✓		✓
Interpretable results!				✓			
Integration of existing knowledge							
Integration with standard software	✓	✓	✓				
Visualization	✓						
Evaluation methods			✓			✓	✓
Reproducible results	✓	✓	✓			✓	✓
Small number of user-defined parameters	✓	✓	✓	✓	✓	✓	✓

3. TRANSITIVITY CLUSTERING

This chapter will describe Transitivity Clustering (TC), a universal clustering approach based on graph modification. The underlying model is the Weighted Transitive Graph Projection (WTGP). This model defines a similarity graph $G = (V, E)$, as an undirected simple graph. An edge is drawn between two objects of V if their similarity exceeds a given similarity threshold, the density parameter of this approach. Afterwards edge modifications, adding and deleting edges, are performed to make this graph transitive, i.e. every connected component is completely connected. The obtained clusters are the cliques of the resulting graph. The NP-hardness of this problem has already been proven in 1986 by Křivánek and Morávek [53] and a first approximation algorithm, the Cluster Affinity Search Technique (CAST) was introduced in 1999 by Ben-dor et al. [18]. Ben-dor et al. predicted co-expressed genes of microarray experiments using the CAST algorithm.

Here, the Weighted Transitive Graph Projection Problem (WTGPP) will be presented and useful attributes of TC will be proven. The mean similarity within one cluster, for instance, is always above the chosen threshold. On the other hand, the mean similarity between two clusters is always below the threshold. If knowledge about the similarity function is available, this eases finding a meaningful threshold. A drawback of clustering data by using this approach is, as in many other approaches, the necessity to solve a NP-complete problem. A helpful property of the underlying model to overcome this problem is that one can divide the problem into much smaller problems. The initial graph can be divided into subgraphs connected by edges above the threshold, which are subsequently solved independently. This may reduce the complexity enormously, allowing to find solutions in a reasonable time and allowing for parallelization to further decrease runtime in practice.

Moreover, this chapter will describe how the WTGPP can be modified to integrate existing knowledge. Elements can, for example, be forced to belong to the same cluster or be forbidden to be assigned to the same group. This can be clusters of objects which are experimentally validated or pairs of objects that exceed a second, much higher, threshold. Although not used within experiments included in this work, one can change the costs for edge modifications. Deleting an edge can be more penalized then adding edges to rather produce larger clusters. In this chapter it will further be shown that clustering with increasing thresholds does not lead to an hierarchical clustering. Two approaches will be presented to overcome this problem, thus allowing to use TC as an alternative for commonly used hierarchical clustering methods, while still taking advantage of the attributes of TC. Furthermore, methods to predict overlaps between clusters are described, which widens the applicability of TC even more.

The chapter concludes with a description of existing algorithms that solve a WTGPP.

3.1 Transitive graph projection

This section introduces the graph modification problem that is used as the model for the clustering approach of this thesis, the WTGPP. To begin with, a related problem, the Transitive Graph Projection Problem (TGPP), will be described, and later extended to the WTGPP.

For an undirected graph $G = (V, E)$ the TGPP aims to find that transitive graph $G' = (V, E')$, with the least amount of edge modifications (additions/deletions) from G to G'. Formally it is defined as follows.

Problem 3.1 (Transitive graph projection). Given an undirected simple graph $G = (V, E)$, find a transitive graph $G' = (V, E')$, such that the edge modifications to derive G' from G are minimal. An edge modification is either an addition, if the edge was not present in E, but exists in E', or a deletion respectively. The costs for adding or removing an edge is 1, and consequently the costs to get G' from G is:

$$\text{costs}(G \to G') = \underbrace{|E \setminus E'|}_{\text{deletion cost}} + \underbrace{|E' \setminus E|}_{\text{addition cost}}$$

Zahn et al. [80] first mentioned the TGPP in 1964. Its NP-hardness was later proven by Křivánek and Morávek [53] and again by Shamir et al. [68]. A Fixed-Parameter (FP) algorithm was first introduced for this problem by Gramm et al. [40]. Furthermore Grötschel and Wakabayashi formulated the problem as an Integer Linear Program in [41]. In the literature it is also referred to as "Cluster Editing", "Correlation Clustering", or "MinDisAgreement" problem.

The TGPP may be used for clustering data but lacks for precision, since no difference can be made between the different grades of similarity between two objects. To overcome this restriction the model can be extended by adding such weights, and changing the modification costs for each edge depending on the similarity between the adjacent nodes. The WTGPP will serve as model for the clustering approach in this thesis. It extends the TGPP, by adding weights to each pair of nodes and using a threshold to determine whether an edge exists or not. In this case, the edge changes are not treated equally but depend on the weights and the threshold. Each pair of objects is assigned a modification cost, which is the absolute of the difference of the weight and the threshold. Thus it is more expensive to add/delete edges with weight far away from the threshold.

Problem 3.2 (Weighted transitive graph projection). Given a set of objects V, a threshold $t \in \mathbb{R}$, and a pairwise similarity function $\text{sim}\colon \binom{V}{2} \to \mathbb{R}$, the graph G is defined as:

$$G = (V, E);\ E = \left\{ uv \in \binom{V}{2}; \text{sim}(uv) > t \right\}$$

The goal of the weighted transitive graph projection problem is to determine a transitive

3.1. TRANSITIVE GRAPH PROJECTION

graph $G' = (V, E')$, such that there exists no other transitive graph $G'' = (V, E'')$, with $\text{cost}(G \rightarrow G'') < \text{cost}(G \rightarrow G')$. Hereby the modification costs are defined as:

$$\text{cost}(G \rightarrow G') = \underbrace{\sum_{uv \in E \setminus E'} |\text{sim}(uv) - t|}_{\text{deletion cost}} + \underbrace{\sum_{uv \in E' \setminus E} |\text{sim}(uv) - t|}_{\text{addition cost}}$$

Note that more than one solution for a given problem instance may exist, but this case almost never occurs in practice, if the similarity function is diverse and real-valued.

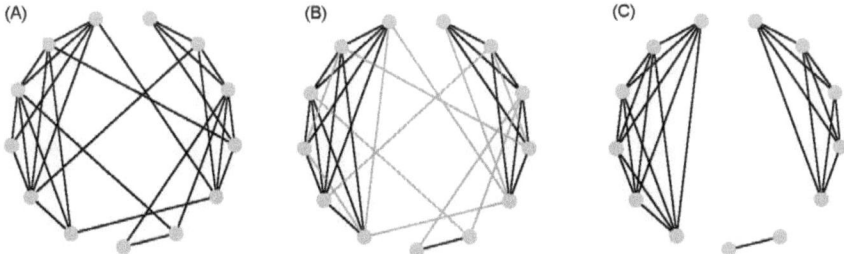

Fig. 3.1: Illustration of the (W)TGPP. (A) shows the similarity graph $G = (V, E)$ with edges E whose value exceed the threshold. (B) shows a putative solution, where red edges have to be removed and green edges are added. (C) shows the transitive graph $G' = (V, E')$ where each connected component is a clique

Since the WTGPP is a special case of the TGPP it is a straight forward task to show its NP-hardness.

Lemma 3.3. *The weighted transitive graph projection problem is NP-hard.*

Proof. Each TGPP can be formulated as a WTGPP problem by setting the similarity between two existing edges to 1, between all other nodes to -1, and the threshold to 0. Naturally a solution for this problem is the solution for the corresponding TGPP. □

In 2003, Charikar *et al.* [27] presented a proof that the WTGPP, which they refer to as weighted MinDisAgree, is APX-hard[1]. Although this shows that it is time consuming to solve a WTGPP one may take advantage of the following attribute, which reduces the complexity enormously.

Lemma 3.4. *Given a set of objects V, a threshold $t \in \mathbb{R}$, and a pairwise similarity function* $\text{sim}: \binom{V}{2} \rightarrow \mathbb{R}$ *and a graph $G = (V, E)$ as defined in the WTGPP, it is sufficient to solve the WTGPP of the connected components $G_1, ..., G_m$ of G, i.e. if $G'_1, ..., G'_m$ are solutions for the WTGPP of the connected components of G, $G' = \bigcup_{i=1}^{m} G'_i$ is a solution for the WTGPP of G.*

---
[1] An optimization problem is in APX if there exists a polynomial-time approximation algorithm for a certain fixed error rate. It is APX-hard if any problem in APX can be reduced to this problem. For problems of that class, no polynomial-time approximation algorithm exists for each fixed error rate, unless P=NP

Proof. In order to prove this lemma it is sufficient to show that there exists no solution of a WTGPP with cliques that intersect with multiple connected components of the similarity graph G. This will be done by assuming that such a solution exists and leading this to a contradiction:

Let G' be a solution of a WTGPP with cliques that contain objects from different connected components of G. Edges in G' between objects of different connected components of G correspond to a similarity smaller than the threshold t due to the definition of the similarity graph. Hence, deleting them results in a decrease in costs. Furthermore, deleting all edges between a subset of nodes of a clique and all other nodes of that clique leads to two disjoint cliques, since all nodes within the two sets are still connected. Consequently, splitting the cliques in G' into cliques that have no intersection with two different connected components of G reduces the costs and still respects the transitivity rule. It is in turn a transitive graph with lower costs than G'. This is a contradiction to the minimal-cost criteria of a solution of a WTGPP. □

3.2 Data partitioning by using weighted transitive graph projection

The optimal solution for the WTGPP is an undirected graph $G' = (V, E')$ following the transitivity rule, i.e. all connected components are cliques. The clusters induced by this graph are exactly these cliques. The similarity threshold serves as the density parameter for this approach and hence defines the number of clusters and their sizes. While the WTGPP is NP-hard to solve and some applications may induce problem instances of large size, the main approach presented here uses heuristic methods. In TransClust, a clustering environment based on TC, a combination of heuristic and exact methods are applied to find a close to optimal solution for each problem in reasonable time. Currently available algorithms to solve the WTGPP are described in Section 3.4. The clustering framework TransClust, which integrates most of these algorithms, will be presented in the next chapter.

Advantages of TC over other approaches are its flexibility and the intuitive density parameter. The similarity threshold directly corresponds to the chosen similarity function, and by choosing such threshold, it is defined what to consider as "similar enough". Only some changes, the adding and deleting operations of the WTGPP, are necessary to detect outliers and produce homogeneous clusters. Edges of elements whose similarity is close to the threshold are more likely to be changed, since the costs for the modifications are rather low. The following property of a partitional clustering with TC gives an impression of the clustering results and thus helps specify an appropriate threshold.

Lemma 3.5. *Let $C = \{C_1, ..., C_m\}$ be the cliques/clusters of a solution $G' = (V, E')$ for a given WTGPP with threshold t and similarity function sim.*

(i) The mean similarity between an element u and all other elements of its clique C_u is greater than or equal to the threshold t for all elements $u \in V$.

(ii) The mean similarity between all elements of one cluster C_i is greater than or equal to t for all cliques $C_i \in C$.

3.2. DATA PARTITIONING BY USING WEIGHTED TRANSITIVE GRAPH PROJECTION

Proof. (ii) is a direct consequence of (i). To prove (i), the negative proposition is assumed and lead to a contradiction. Let u be an element of the cluster C_i of size $|C_i| \geq 2$. Assume the mean similarity between u and all other elements of C_i is below t:

$$\text{meansim}(u, C_i) = \frac{1}{|C_i| - 1} \sum_{v \in C_i \setminus \{u\}} \text{sim}(uv) < t$$

$$\Leftrightarrow \sum_{v \in C_i \setminus \{u\}} \text{sim}(uv) < t \cdot (|C_i| - 1)$$

$C' = \{C_1, ..., C_i \setminus \{u\}, ..., C_m, \{u\}\}$ is a decomposition of the elements into cliques and hence a putative solution for the underlying WTGPP. The costs for C' can be calculated by using the costs that appear to build C and adding all costs to remove edges between u and C_i. Note that these additional costs may be negative for edges that did not exist in the initial graph and had to be added to create C_i. Using the assumption that the mean similarity between u and all elements of C_i is below t, the cost difference between C and C' is consequently:

$$\sum_{v \in C_i \setminus \{u\}} (\text{sim}(uv) - t) = \left(\sum_{v \in C_i \setminus \{u\}} \text{sim}(uv) \right) - (t \cdot (|C_i| - 1)) < 0$$

This is a contradiction to the assumptions that C is a solution for the WTGPP, since there exists a decomposition into cliques with lower costs. □

A statement about the average similarity between an object and all the objects of a foreign cluster is not possible. The following example illustrates that one element might have a mean similarity above the threshold to all elements of a different cluster.

Let $V = \{a, b, c\}$ be the elements of interest. Let the similarity between these elements be $\text{sim}(ab) = 0.5$, $\text{sim}(ac) = 0$, and $\text{sim}(bc) = 1$. For a threshold $t = 0.4$ the clustering obtained by solving the corresponding WTGPP is $C = \{C_1, C_2\} = \{\{a\}, \{b, c\}\}$. The mean similarity between objects within one cluster is obviously above the threshold and the mean similarity between these clusters is below the threshold:

$$\frac{\text{sim}(ab) + \text{sim}(ac)}{2} = 0.25 < t$$

The mean similarity between b and a, which is one element from one cluster and all elements from the other, is 0.5 and hence above the treshold.

It is possible though to make a statement about the average similarity between two clusters.

Lemma 3.6. *Let $C = \{C_1, ..., C_m\}$ be the cliques of a solution for a given WTGPP with threshold t and similarity function sim. The mean similarity between two cliques C_i and C_j is below the threshold for all $1 \leq i < j \leq m$.*

Proof. Again the proof for this lemma is done by assuming the negated proposition and leading it to a contradiction. Let C_i and C_j with $i \neq j$ be cliques with average similarity above the threshold t. The decomposition of the objects into cliques $C' = (C \setminus \{C_i, C_j\}) \cup \{C_i \cup C_j\}$ is a putative solution for the WTGPP. The costs for C' can again be calculated using the costs for C and adding all costs for adding the connective edges between C_i and C_j:

$$\text{costs}(C') = \text{costs}(C) + \sum_{u \in C_i} \sum_{v \in C_j} (-\text{sim}(uv) + t)$$

In order to see a contradiction to the assumption that C is a solution for the WTGPP all that remains is to show that the second term is below zero. This can be derived from the initial assumption that the average similarity between C_i and C_j is above the threshold:

$$\text{meansim}(C_i, C_j) = \frac{1}{|C_i| \cdot |C_j|} \sum_{u \in C_i} \sum_{v \in C_j} \text{sim}(uv) > t$$

$$\Leftrightarrow \left(\sum_{u \in C_i} \sum_{v \in C_j} \text{sim}(uv) \right) - t \cdot (|C_i| \cdot |C_j|) > 0$$

$$\Leftrightarrow \sum_{u \in C_i} \sum_{v \in C_j} (\text{sim}(uv) - t) > 0$$

$$\Leftrightarrow \sum_{u \in C_i} \sum_{v \in C_j} (-\text{sim}(uv) + t) < 0$$

$\square$

3.3 Extensions

In order to improve the clustering results, one can modify the WTGPP. One option is to include existing knowledge. Objects, where it is known that they belong to the same cluster can be set to be inseparable for instance, or a second threshold may specify a limit above which two elements are also forced to be in one cluster. Another extension of TC is to compute a hierarchal or an overlapping clustering.

3.3.1 Upper and lower bounds

In some cases it is useful to force some elements to be in one cluster. If two elements are more similar than a second threshold t_u (upper bound), they can be considered as "similar enough" to be inseparable. For instance, proteins whose best bidirectional BLAST E-value is below 10^{-200} can be forced to belong to the same cluster. The following is a description of how such a restriction is defined formally. The integration of known assignments works in a similar way to the modification with an upper bound. Hence, only the modification with upper and lower bound are described here.

Given a set of objects V, a pairwise similarity function sim: $\binom{V}{2} \to \mathbb{R}$ and a threshold $t \in \mathbb{R}$, a WTGPP can be formulated. Using an upper bound t_u, all elements that are connected via a path that consist only of edges with similarity above t_u are assigned to groups. These groups are the connected components $\hat{C} = \{\hat{C}_1, ..., \hat{C}_m\}$ of the graph $G' = (V, E')$ with $E' =$

3.3. EXTENSIONS

$\{uv \in \binom{V}{2}; \text{sim}(uv) > t_u\}$ and serve as new objects for the subsequent clustering. Hence, they can no longer be separated. A group may contain edges below t_u or even below the original threshold t. Further note, that a group $\hat{C}_i$ may consist of only one object. Afterwards the problem of clustering these groups is defined again as WTGPP using a threshold of 0. The new similarity graph $\hat{G} = (\hat{C}, E_{\hat{C}})$ is consequently defined as:

$$E_{\hat{C}} = \left\{ CC' \in \binom{\hat{C}}{2}; \hat{\text{sim}}(CC') > 0 \right\}$$

where $\hat{\text{sim}}$ is the similarity function between the new objects. $\hat{\text{sim}}$ is defined for two groups $C, C' \in \hat{C}$ as the sum of addition/deletion costs (according to the original WTGPP) between all elements of C and all elements of C', where deletion costs are subtracted:

$$\hat{\text{sim}}(CC') := \sum_{u \in C} \sum_{v \in C'} (\text{sim}(uv) - t)$$

This "new" problem is closely related to the original formulation and can even lead to the same results, though not guaranteed. It can be shown easily (see Figure 3.2 as an example), that solving the restricted problem may lead to a partitioning of the data with higher costs according to the original WTGPP.

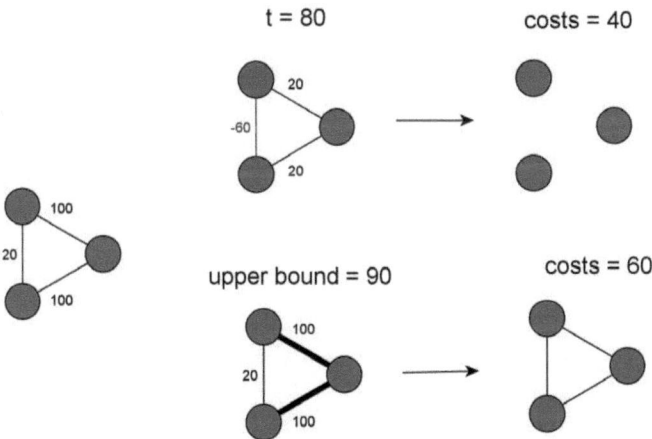

Fig. 3.2: Counter example, that the WTGPP with upper bound lead to higher costs. Using a threshold $t = 80$ and an upper bound $t_u = 90$. Without the upper bound all nodes are separated with costs 40 to remove the two remaining edges, and with upper bound all nodes are forced to be in one cluster, leading to costs 60 for adding the missing edge.

This operation may also be used as a heuristic to solve the initial WTGPP. The costs for the corresponding partitioning of G, given a solution $\hat{G}'$ for $\hat{G}$, is the sum of the costs for creating this solution and the costs to build the inseparable groups with t_u. These additional costs are on the one hand the costs for adding all non-existing edges within a group of inseparable elements and on the other hand costs for defining the edges between two groups or a group and another object. If an edge between two groups C and C' exists, the costs for creating this edge is the sum of costs for adding all non-existing edges between objects of C and objects of C'. If an edge does not exist between C and C, the additional costs are the sum of costs for removing all edges between objects of C and C'. These costs can be stored during the construction of the graph $\hat{G}$.

A second option to influence the results with additional knowledge, is to set some edges to forbidden, meaning not allowing the adjacent nodes to be in one cluster. As in the first restriction this would not lead to an optimal solution of the WTGPP, but may reflect reality more precisely. An example for this can be found in the application chapter, Section 6.1. It is possible to set a lower bound t_l. Each similarity smaller than t_l is set to $-\infty$. Note that combining the two methods may lead to contradiction, since two elements with a similarity below t_l may still be part of the same cluster due to a path of edges with similarity above the upper bound. In the work presented here, the upper bound will be applied first.

3.3.2 Building hierarchies

Due to the nature of the underlying graph modification problem, TC is a partitional clustering algorithm and, even with strictly ascending or descending thresholds, no hierarchy is achieved (Figure 3.3 illustrates such an example). The question arises, in which cases no hierarchical clustering occurs and why. TC depends on the threshold and on the number of elements that are similar to each other. If objects from a large set S are similar to one object x, x might be assigned to S even if it has higher similarities to other elements. The following example illustrates this effect:

Let V be the set of objects and assume there exist two disjoint groups $U, W \in V$. The similarity of all elements within U and W is high, but low between elements of U and elements of W. For simplification let the similarity within a group be 1 and between the groups be 0. Furthermore, assume there exists one element $x \in V \setminus (U \cup W)$, which has similarity to all elements of U of 0.5 and a similarity to all elements of W of 1. Choosing a threshold above 0.5 clearly assigns x to W, but if the threshold is below 0.5 the assignment of x depends on the size of U and W.

To assign x to U the costs for removing all edges between x and W must be lower than removing all edges between x and U. In this example, one can ignore the edge additions,

3.3. EXTENSIONS

since all edges are above the threshold.

$$\sum_{w \in W} \text{sim}(xw) - t \leq \sum_{u \in U} \text{sim}(xu) - t$$
$$\Leftrightarrow |W| \cdot (1-t) \leq |U| \cdot (0.5-t)$$
$$\Leftrightarrow |W| \cdot \frac{1-t}{0.5-t} \leq |U|$$

For a threshold $t = 0.25$, x is assigned to U if the number of elements of U is more than three times the number of elements in W:

$$|W| \cdot \frac{1-t}{0.5-t} = |W| \cdot \frac{0.75}{0.25} = |W| \cdot 3 \leq |U|$$

It is exactly this effect that is responsible for the WTGP to be non-hierarchical. A similar example can be found in Figure 3.3, where x is similar to u and v, but more similar to w. For a small threshold of 4, u and v combined are more similar to the crucial node x. For a higher threshold of 7 the highest similarity between x and w is important in determining the assignment of x.

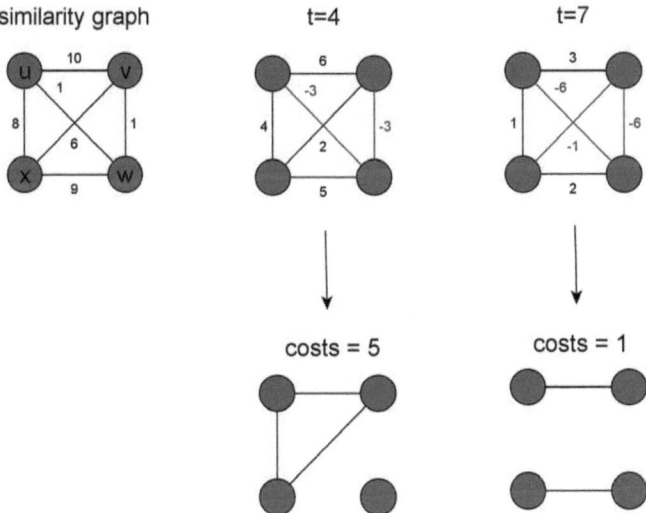

Fig. 3.3: This illustrated example shows, that the WTGP does not produce an hierarchical structure for ascending thresholds. Using thresholds $t = 4$ and $t = 7$ leads to different solutions as expected, but one can see that the clusters for the higher threshold $t = 7$ are not subclusters of the groups obtained with threshold $t = 4$

To avoid such behavior, one may either force elements with high similarity to be in one cluster, or force an hierarchical structure as in one of the two options; a bottom-up and a top-down method. Both methods iterate through a list of thresholds, ascending for the top-down approach and descending otherwise. While the top-down method always uses the obtained clusters from the previously used threshold to create the input graphs for the current parameter, the bottom-up method merges all nodes of the previous clusters and calculates the new similarity as sum of similarities between nodes from different cluster. If the bottom-up approach is chosen, elements with high similarity are assigned to the same cluster in early iterations, and are not allowed to be split afterwards. On the other hand, the top-down approach prefers to first build big groups of elements with low similarity that will be split further in future iterations. With both methods one does not lose the advantages of the WTGP.

Top-down

As described above, the top-down approach needs a list of ascending thresholds $t_1, \ldots, t_k$, with $t_i < t_j$ for $1 \leq i < j \leq k$. For t_1, the WTGPP is formulated as usual. Given a solution G' consisting of cliques $C_1, \ldots, C_m$, each clique defines a graph G_i which serves as similarity graph for the next threshold, i.e. a WTGPP can be formulated for each clique individually using the next threshold t_2 and the graphs G_i, defined as follows:

$$G_i = (C_i, E_i); \ E_i = \{\{u,v\} \in \binom{C_i}{2}; \mathrm{sim}(uv) > t_2\}$$

By iterating over all thresholds and by using the results of the previous solutions to construct the next problem, a hierarchy is created. The mean similarity between elements of one cluster is still above the threshold t_i for the corresponding clustering in iteration i.

Lemma 3.7. *The previously described attributes for similarities between elements and their group as well as between two clusters hold true for clusters obtained after i steps of the top-down hierarchical approach, i.e.*

(i) *The mean similarity between one element and all elements of its cluster is greater or equal to the threshold t_i at iteration i.*

(ii) *The mean similarity between all elements of one cluster obtained in the i-th iteration is greater or equal to the threshold t_i.*

Proof. (ii) is a direct consequence of (i) and hence does not have to be proven independently. As shown in Lemma 3.5, the statements are valid for the first iteration. The results for each iteration i are used to construct a WTGPP. These problems are independent from each other and hence follow the rules defined in Lemma 3.5. □

Unfortunately, it can not be proven that the mean similarity between two clusters is below the threshold. This statement is obviously true within one graph obtained in the i-th iteration

3.3. EXTENSIONS

and clustered with threshold t_{i+1}. It cannot be proven for comparing clusters of two different graphs of the i-th iteration, since those graphs may contain elements with mean similarity to another graph above the threshold t_i and even above t_{i+1} (see example 3.2).

Bottom-up

The second approach for creating a hierarchy using WTGP, starts with a set of small clusters and merges them until either only one cluster is left or the list of thresholds is processed. In contrast to the previous method, the required list of thresholds $t_1, \ldots, t_k$ has to be descending ($t_i > t_j$ for $1 \leq i < j \leq k$). Starting with the highest value t_1, the WTGPP is formulated and solved. The cliques of the solution are the objects for the next threshold t_2. As described in Section 3.3.1, elements can be forced to belong to the same cluster with an upper bound or a pre-specified assignment. In this case all elements of one clique must not be separated anymore. The costs to delete or add an edge between these constructed objects are defined as the sum of differences of the weight and the next threshold t_2 between elements of two groups. If this sum is greater then zero the edge exists and otherwise it does not. Again, this process iterates through the list of thresholds and, thus constructs a hierarchy. It seems to be similar to a standard agglomerative hierarchical clustering, yet still differs. Rather than one element being added to an existing cluster in each step, several may be assigned to the same group, depending on the list of threshold. The clustering obtained for a threshold t_i still guarantees, that the mean similarity within one cluster is above the threshold.

Lemma 3.8. *A partitional clustering obtained in the i-th step of the bottom-up hierarchical clustering has the following attributes:*

(i) The mean similarity between one element and all other elements of its cluster is greater or equal to the threshold t_i at iteration i.

(ii) The mean similarity between two clusters obtained in the i-th iteration is below the threshold t_i.

Proof. (i) Again the statements for the first iteration are true due to Lemma 3.5. Assuming the statements are true for the i-th iteration, what remains is to show that this leads to the correctness for the $i+1$-th iteration.

Solving the WTGPP for the next threshold t_{i+1} guarantees (again from Lemma 3.5) that the mean similarity within one cluster C is above the threshold:

$$\sum_{U \in C} \sum_{V \in C \setminus \{U\}} \operatorname{sim}(UV) > \binom{|C|}{2} \cdot t_{i+1}$$

where the similarity between two elements (clusters for t_i) U and V is, as previously described, defined as:

$$\operatorname{sim}(UV) = \left(\sum_{u \in U} \sum_{v \in V} (\operatorname{sim}(uv) - t_{i+1}) \right) + t_{i+1}$$

This shows that the average inner similarity of clusters derived with t_i in C is above the threshold t_{i+1}. Since it is assumed that the statement is correct for the previous iteration, the remaining similarities, i.e. the average similarities within the clusters obtained using t_i, are above t_i and, since $t_i > t_{i+1}$, also above t_{i+1}. From this follows that the mean similarity between all elements of one cluster C is also above the threshold t_{i+1}

(ii) From Lemma 3.5 it is known that the mean similarity between two clusters C and K of the $(i+1)$-th iteration is below the corresponding threshold:

$$\text{meansim}(CK) = \frac{1}{|C|\cdot|K|} \sum_{U \in C} \sum_{V \in K} (\text{sim}(UV)) < t_{i+1}$$

Together with the definition of similarity between two clusters, this proves that the average similarity between the original objects from C and K is again below the threshold.

$$\sum_{U \in C} \sum_{V \in K} (\text{sim}(UV) - t_{i+1}) < 0$$

$$\Leftrightarrow \sum_{U \in C} \sum_{V \in K} \left(\left(\sum_{u \in U} \sum_{v \in V} (\text{sim}(uv) - t_{i+1}) \right) + t_{i+1} - t_{i+1} \right) < 0$$

$$\Leftrightarrow \frac{1}{((\sum_{U \in C} |U|) \cdot (\sum_{V \in K} |V|))} \cdot \sum_{U \in C} \sum_{V \in K} \left(\sum_{u \in U} \sum_{v \in V} \text{sim}(uv) \right) < t_{i+1}$$

□

These two approaches can be used to create an hierarchical clustering based on the WTGPP. Alternatively it is an option to decrease the runtime of an iterative analysis at the expense of accuracy. It can be used to get a fast approximation about the results before the execution of a more precise method. As described above, the hierarchical clustering methods are a way to control the outcome of the clusters, by preferring either elements with high similarity to be in one cluster (bottom-up) or bigger groups with an overall similarity between the objects (top-down).

3.3.3 Identification of overlaps

This section describes three methods to create an overlapping clustering using WTGP. The first approach creates a fuzzy assignment of the objects to clusters by using the partitional clustering for a certain threshold. Based on the similarity function, each element-cluster pair is assigned a value between zero and one, which is subsequently used to assign those objects to additional clusters whose value exceeds a second threshold. Another method is more related to the initial WTGPP. Again, the WTGPP is solved first for one threshold. Afterwards single elements are assigned to other clusters if this reduces the costs. As in any overlapping clustering, the transitivity rule can not be fulfilled anymore. The last approach has been specifically developed for clustering proteins based on their sequence similarity. Sequences compared using the BLAST algorithm, result in multiple High Scoring Pairs (HSPs) each representing an alignment of subsequences of two protein sequences. While common clustering

3.3. EXTENSIONS

approaches primarily use only the best bidirectional hit, i.e. ignoring the position of the aligned subsequences and multiple HSPs for the same sequences, the method presented here makes use of all this information.

In the following let $C = \{C_1, ..., C_m\}$ be a clustering obtained for a specific threshold by solving the corresponding WTGPP, let $V = \{v_1, ..., v_n\}$ be the objects that were clustered, and let sim: $\binom{V}{2} \to \mathbb{R}$ be the pairwise similarity function.

Overlapping clustering with fuzzy associations

This method assigns to each object v_i and each cluster C_j a value $f_{i,j} = f(v_i, C_j) \in [0, 1]$, such that:

$$\sum_{C_j \in C} f(v_i, C_j) = 1 \text{ ; for all } v_i \in V$$

Therefore, the mean similarity $m_{i,j}$ between every pair of object v_i and cluster C_j is calculated and stored in the matrix $M = (m_{i,j}) \in \mathbb{R}^{n \times m}$:

$$m_{i,j} = \frac{1}{|C_j \setminus v_i|} \sum_{v \in C_j, v \neq v_i} \text{sim}(vv_i)$$

The matrix is transformed into a column stochastic matrix $F = (f_{i,j}) \in [0, 1]^{n \times m}$ to fulfill the criteria as defined above. f is a value that determines how "well" an element fits into a cluster. To get an overlapping clustering a second threshold is necessary. This value $t_2 \in [0, 1]$ influences the number of allowed overlaps. Choosing a low value would lead to many overlaps, while a value close to one would only assign few elements to multiple clusters. The overlapping clustering itself is created by adding an element v_i to an additional cluster C_j if $f_{i,j}$ exceeds t_2. One restriction is that singletons cannot be assigned to any other cluster and that no element can be assigned to singletons.

Adding single objects to multiple clusters

The second approach also takes advantage of a previous clustering obtained by solving a WTGPP. The basic concept is to assign objects to an additional cluster if the internal costs is below zero. By adding a node to an additional cluster and hence all edges between the node and the cluster, the overall costs may be reduced if initially removed edges are re-added again. In this approach, the transitivity rule no longer applies, as with any overlapping clustering. The strength of the overlap cannot be specified as with the previous approach, and only depends on the initially chosen threshold and obtained clustering. Starting with a partitional clustering $C = \{C_1, ..., C_m\}$ each object u can be assigned to an additional cluster C_j if

$$\text{costs}(u, C_j) = \sum_{v \in C_j} (\text{sim}(uv) - t) < 0$$

For the combination u, C_j with smallest costs and $\text{costs}(u, C_j) < 0$, the altered clusters $C' = \{C_1, ..., C_j \cup \{u\}, ..., C_m\}$ replace the previous clusters and the process starts again. These operations are executed until no further improvement is possible, i.e. no assignment of an object to an additional cluster would reduce the internal costs of any cluster. As before, singletons are excluded as well as assignments to singletons.

With both overlapping methods, the mean similarity between an object and all other object of its cluster is above the chosen threshold.

Overlapping clustering of protein sequences

The last overlapping approach has been developed particularly for the task of clustering proteins. A sequence comparison by using the BLAST algorithm serves as input for this approach. The method is motivated by the fact, that proteins in a certain family share domains and do not have to be completely identical. In some cases it occurs that a protein has a subsequence similar to one group and another subsequence similar to a second group. It is difficult to decide which group such a protein should be assigned to. This problem can be solved by allowing overlaps thus assigning it to both clusters. BLAST results are organized in HSPs, each representing an alignment between a subsequence of one protein sequence to a subsequence of another. Commonly the best bidirectional E-value or the normalized score of these alignments are used as pairwise similarity. Instead of clustering whole protein sequences, only the in the BLAST file occurring subsequences are taken into account here, i.e. the subsequences of a protein that have a high similarity to subsequences of other proteins. These sequences are initially clustered to identify regions within one protein sequence that can also be found in sequences from other proteins. For each occurring two subsequences s_1 and s_2 of the same protein p of length $|p|$ the following value indicates how similar they are:

$$\text{sim}(s_1, s_2) = \frac{\sum_{i=1}^{|p|} \min\{I_{s_1}(i), I_{s_2}(i)\}}{|p|}$$

where $I_s : \{1, ..., |p|\} \to \{0, 1\}$ is a function that indicates if the subsequence s covers the i-th position of p. s_1 and s_2 are treated as the same subsequence of p if $\text{sim}(s_1, s_2) > t$, where $t \in [0, 1]$ is a fixed threshold. Such groups of "almost identical" subsequences are the objects in a subsequent cluster analysis using TC. An overlapping clustering can be obtained by identifying the subsequences again with their proteins. With this method it might be possible to identify connecting proteins which have a similar subsequence to one group and another one to a different group. Although not verified yet those results may help detecting groups of domains that are representatives for functionally related groups of proteins.

3.4 Algorithms solving the weighted transitive graph projection problem

3.4.1 Fixed parameter branch and bound strategy

A FP strategy for the TGPP has first been introduced by Gramm et al. [39]. Later, Böcker et al. [23] extended this approach to the WTGPP and added several improvements to decrease the runtime. The current implementation of the subsequently presented FP approach, called PEACE, is available as web server at http://bio.informatik.uni-jena.de/peace/.

Essentially, a branch and bound strategy is used with an increasing parameter k as an upper bound for the modification costs. The trick is to restrict the search space to those edges that contradict with the transitivity rule. Note that changing one edge within a so-called conflict triple (three nodes and two edges) is sufficient to resolve the contradiction to the transitivity rule. Thus, for one edge the search tree branches into two subtrees: To keep an edge means to assign the connected objects to the same cluster, while removing an edge means to separate them into different clusters. The modification costs are calculated for each step and the branching stops if a certain k is exceeded. Keeping an edge is processed by merging the two adjacent nodes and thus reduce the number of nodes for this subtree by one. Each edge is processed at most once, which guarantees to find a solution with costs smaller than k, if one exists. If no solution was found for a given k, the algorithm is restarted with an increased value of k. This is repeated until a solution is found. To improve runtime additional reduction rules are implemented. These check in an efficient way for a given edge whether it is possible to find a solution where the edge is kept, or removed. If no solution can exist the edge is removed or kept respectively. The minimal cost for removing the edge uv are:

$$\text{minremovalcost}(uv) = \sum_{w \in N(u) \cap N(v)} \min\{\text{sim}(uw), \text{sim}(vw)\}$$

and the minimal cost for keeping the edge uv are:

$$\text{minkeepingcost}(uv) = \sum_{w \in N(u) \Delta N(v)} \min\{|\text{sim}(uw)|, |\text{sim}(vw)|\}$$

where $A \Delta B$ denotes the symmetric set difference of two sets A and B, and $N(u)$ is the set of neighbors of u. Note that the worst case running time of the algorithm is still exponential but for problem instances where only few edge modifications are necessary, the exact solutions can be found early. In practice, at least for protein sequence clustering, for connected components of size 200 nodes or more, the algorithm may not be able to find a solution in reasonable time (less than 24 hours on a standard desktop computer).

3.4.2 FORCE

The FORCE approach [79] is the predecessor of the software TransClust, which is described in the next chapter. It is a heuristic whose main concept is to arrange the nodes of the graph on a plane in such way, that the layout reflects the similarity. Afterwards a single linkage approach is used to determine the clusters. FORCE is organized in three steps; (1) the layout step, (2) the partitioning step, and (3) the post-processing step. In the layout step a graph layout algorithm similar to the force-based algorithm by Fruchterman and Reingold [35] is applied to organize the nodes on the plane. Since this algorithm is similar to the layout algorithm in TransClust, a description can be found in Section 4.2.1 and Algorithm 1 on page 53. The partitioning step applies a simplified version of the single linkage clustering algorithm, where for a fixed list of distances $d_1, \ldots, d_n$ all nodes with distance smaller than d_i are assigned to one cluster and the costs are calculated. The solution with smallest cost is then passed to the last step, the post-processing step. The two post-processing methods "merge" and "rearrange single nodes" are used here. Descriptions of these methods can also be found in Section 4.2.4, since they are also implemented in TransClust.

3.4.3 A greedy approach

A heuristic greedy approach has been developed by Marcel Martin [65] for this task. The main idea is to correctly guess the set of edges to remove from the input graph G and afterwards take the transitive closure of the remaining graph as solution. To determine which edges are good candidates for removal, a score for each edge removal is defined. Let $C(G)$ denote the set of conflict triples, i.e. the set of triples $uvw \in \binom{V}{3}$ that contradict with the transitivity rule. The deviation from transitivity is then defined as:

$$D(G, s) = \sum_{uvw \in C(G)} \min\{|\operatorname{sim}(uv)|, |\operatorname{sim}(uw)|, |\operatorname{sim}(vw)|\}$$

Finally, the transitivity improvement and thus a score for an edge removal is defined as:

$$\operatorname{score}(uv) = D(G, \operatorname{sim}) - D(G', \operatorname{sim}') - \operatorname{sim}(uv)$$

where sim is the similarity function and $G'(V, E \setminus uv)$ is the graph G with edge uv removed. sim' equals sim except for the edge uv, where it is set to $\operatorname{sim}'(uv) = -\infty$. The greedy algorithm searches for the best deletions according to the score function until G is split into two connected components G_1 and G_2. All edge removals that did not contribute to the splitting of G are re-added to G_1 and G_2. Subsequently the algorithm is applied on the subgraphs recursively until a subgraph is a clique or the cost for splitting the graph are higher than those for the transitive closure.

3.4.4 Integer Linear Programming

An Integer Linear Programming (ILP) formulation was first provided by Grötschel and Wakabayashi [41]. In the following, let x be a decision vector with $x_{uv} = 1$ if the edge uv is part of the solution and $x_{uv} = 0$ otherwise. For simplification, let $x_{v_i v_j}$ be denoted as $x_{i,j}$ for all $v_i v_j \in \binom{V}{2}$. The WTGPP can now be formulated as

3.4. ALGORITHMS SOLVING THE WTGPP

Problem 3.9 (ILP formulation for WTGPP).

$$\begin{aligned}
\text{minimize} \quad & \sum_{uv \in E} \text{sim}(uv) - \sum_{1 \leq i < j \leq n} \text{sim}(v_i v_j) x_{i,j} \\
\text{subject to} \quad & + x_{i,j} + x_{j,k} - x_{i,k} \leq 1 && \text{for all } 1 \leq i < j < k \leq n \\
& + x_{i,j} - x_{j,k} + x_{i,k} \leq 1 && \text{for all } 1 \leq i < j < k \leq n \\
& - x_{i,j} + x_{j,k} + x_{i,k} \leq 1 && \text{for all } 1 \leq i < j < k \leq n \\
& x_{i,j} \in \{0, 1\} && \text{for all } 1 \leq i < j \leq n
\end{aligned}$$

Grötschel and Wakabayashi proposed a cutting-plane approach to solve this problem, which has been implemented and improved by Böcker et al. [24]. It has been shown in a comparison between the ILP and the FP approaches, that the two methods have similar runtime in practice.

3.4.5 Cluster Affinity Search Technique

The CAST algorithm by Ben-dor et al. [18] uses a fast technique to predict clusters and thus present putatively 'good' solutions for the WTGPP. The algorithm opens a cluster C_{open} with an arbitrary node. For each element $u \in V$ the affinity to this cluster is calculated as the sum of the similarities between u and all elements of C_{open}:

$$\text{aff}(u, C_{\text{open}}) = \sum_{v \in C_{\text{open}}} \text{sim}(uv)$$

The algorithm now alternates between two methods, ADD and REMOVE:

ADD The node with the highest affinity to C_{open} is added to it, if $\text{aff}(u, C_{\text{open}}) \geq t \cdot |C_{\text{open}}|$. This is repeated until no such node exists or all nodes are assigned to C_{open}. After each assignment, the new affinities of all remaining elements to C_{open} are updated.

REMOVE The node u with lowest affinity to C_{open} is removed from it, if $\text{aff}(u, C_{\text{open}}) < t \cdot |C_{\text{open}}|$. As in ADD this is repeated until no such element is found or C_{open} contains only one element. After each removal, the new affinities of all remaining elements to C_{open} are updated.

When no changes occur in both of these methods, C_{open} is defined as one cluster and another node is picked out of the remaining set $V \setminus C_{\text{open}}$. The process is repeated until every node is assigned to one cluster.

CAST already implemented a post-processing method similar to that in TransClust, which moves one node to a different cluster if this would reduce the costs. This operation is performed until no further movements would reduce the costs, or a user defined maximal number of movements is reached.

4. THE TRANSITIVITY CLUSTERING FRAMEWORK TRANSCLUST

The clustering framework TransClust has been developed as an implementation of Transitivity Clustering (TC). It combines heuristic and exact approaches to optimize quality and runtime. A first implementation and hence predecessor of TransClust, the Force-Based Cluster Editing (FORCE) program, mainly uses a layout based heuristic, inspired by physical forces between objects [79]. An improved version of this algorithm serves as main heuristic approach in TransClust. In addition to this, the Cluster Affinity Search Technique (CAST) algorithm is integrated as well as an exact algorithm. While the CAST heuristic is fast, but less accurate, the exact Fixed-Parameter (FP) strategy, developed by Böcker et al. [23], can be applied on small problem instances only. Best results are obtained by combining all these approaches. A fast approximation for an upper bound leads to decreased runtime of the exact algorithm, and different problem instances may be better solved by using different heuristic methods. The current implementation of the FP strategy is also used later to evaluate the heuristic clustering process of TransClust.

Existing knowledge can be used in TransClust to detect a meaningful similarity threshold. If, for example, a clustering of parts of the used data set is known, results for varying thresholds may be used to see which parameter identifies these clusters best. Additionally, a visual representation supports the data analysis. For this reason the developed clustering framework TransClust has been integrated into the network visualization and analysis software Cytoscape [69]. Together with the clustering methods of TransClust various follow up analysis steps may also be performed here. For instance, the cluster size distribution may be visualized, or a similarity function can be evaluated together with a gold standard assignment by plotting the similarities between two clusters and those within a cluster. All these methods may support a scientist by detecting a meaningful threshold for the clustering problem of interest.

TransClust can be applied in three different ways: (1) as a web application for small problem instances, (2) as a standalone application with either graphical user interface or via commandline, and (3) as a plugin for the network analysis software Cytoscape.

This chapter provides a detailed description of the TransClust framework. Figure 4.1 illustrates the structure of TransClust. The used data formats are presented on the left side and linked to the parts of TransClust, where they are used. The middle briefly summarizes the different clustering options and the right part illustrates the three ways to execute TransClust.

48   4. THE TRANSITIVITY CLUSTERING FRAMEWORK TRANSCLUST

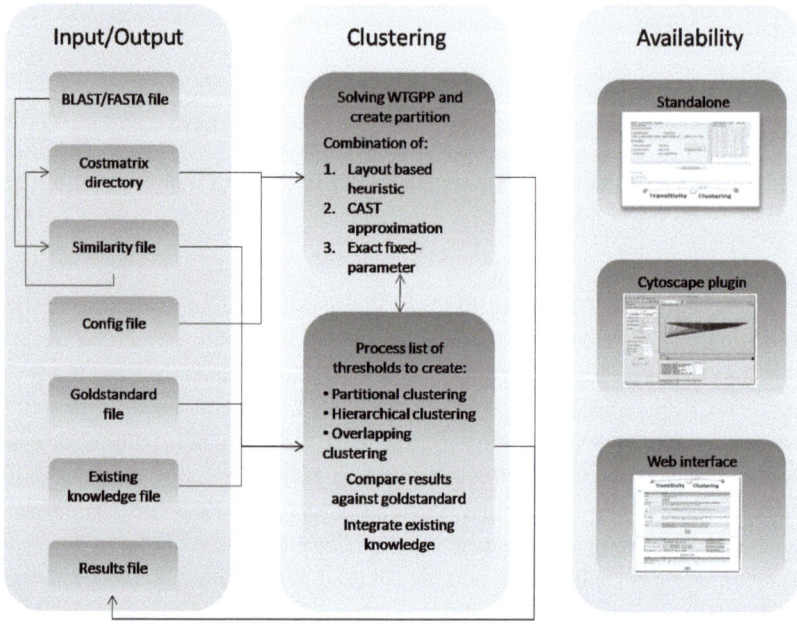

Fig. 4.1: Illustration of the TransClust program structure.

4.1 Data import

TransClust accepts several data formats as input for a subsequent cluster analysis. One of the first application of TC was the clustering of protein sequences. One accepted format is thus a BLAST and a corresponding FASTA file. From these files a similarity file is created. Such a similarity file may also be used as input for any other clustering task, where a pairwise similarity is given. To reduce the required space, an individual Weighted Transitive Graph Projection Problem (WTGPP) can be stored as costmatrix file. In such file, the names of the objects appear only once, and only one direction of the symmetric pairwise similarity function is stored. Further data formats are: a gold standard file, as reference for a comparison, a config file to store configurations of the used heuristics, the results file containing all information about the resulting cluster analysis, and a file containing existing information such as known assignments for instance.

A description and examples for all data formats can be found in the Appendix section B. An overview of the import process of TransClust is illustrated in Figure 4.2.

4.2 Clustering methods

TransClust integrates several algorithms solving a WTGPP. Exact and heuristic methods are combined in an efficient way to minimize the trade-off between runtime and quality. This section

4.2. CLUSTERING METHODS

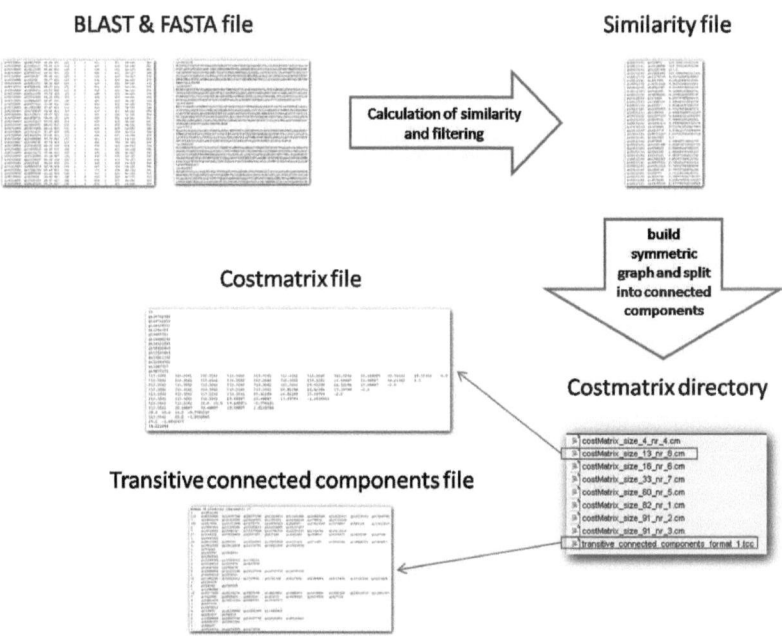

Fig. 4.2: Model of the data import process. For clustering sequences a BLAST and FASTA file is converted into a similarity file. The similarity file can be used to either calculate clusters for varying thresholds directly or to compute costmatrices. Together with the costmatrices a file is created that contains every transitive connected components.

describes all the clustering methods used and how they are combined. It also includes methods that were developed for TransClust, but have proven to be less efficient than the currently used alternatives. However, the integration of novel algorithms for parts of the clustering pipeline, shows that TransClust is easy to extend and, moreover, that it is already optimized for runtime and quality aspects.

Figure 4.3 depicts the architecture of the clustering process. Since it is sufficient to solve a WTGPP for each connected component individually, the picture illustrates the workflow in the example of one intransitive connected component. First, the fast CAST is applied to preprocess the data. This initial clustering leads to a fast approximation of the required costs for this problem. TransClust benefits in two ways from this approximations. For every two nodes the minimum costs for removing the connecting edge can be calculated. If these costs are higher then the costs obtained with CAST these two nodes can be merged to one node to reduce the problem complexity. The second benefit of the approximated costs appears while applying the exact FP algorithm. The runtime of this algorithm can be drastically reduced by guessing a good upper bound. Since the FP algorithm has still exponential runtime its usage

50   4. THE TRANSITIVITY CLUSTERING FRAMEWORK TRANSCLUST

is limited in TransClust. Only instances of a certain size are calculated exactly, and even then the calculation is limited to a certain time. The main algorithm in TransClust is a layout-based heuristic, which is slower then the CAST method but more accurate. Both results, those from CAST and from the layout-based heuristic are transfered to the post-processing phase. After fine-tuning the clustering new connected components are created for each cluster and the procedure starts again. If this recursion does not lead to any improvement, the best result of the two heuristics is reported. Note that TransClust is capable of computing results for different connected components in parallel. One can specify the number of CPUs to be used. This might not decrease running time for an individual large component, but will in most cases speed up the overall clustering procedure.

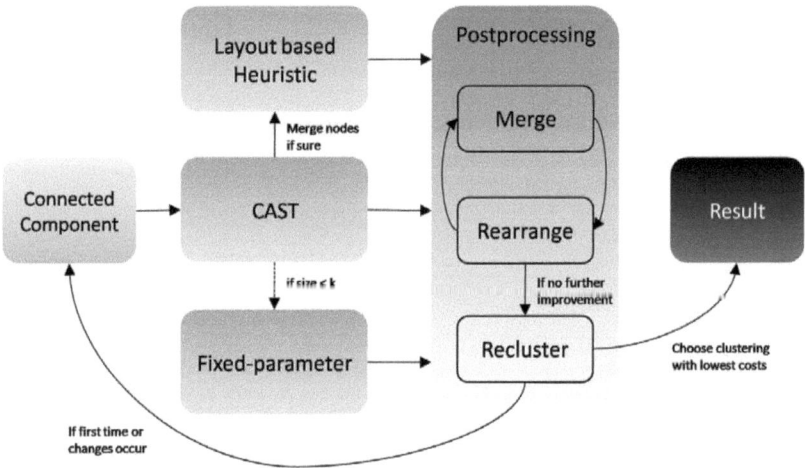

Fig. 4.3: Model of the clustering process. First CAST is applied to produce a quick approximation of costs. All nodes for which deleting the edge would lead to more costs than the pre-calculated costs are merged. Next, depending on the size, either the exact FP algorithm or the layout-based heuristic is applied. The resulting clusters are then post-processed and each cluster is subsequently clustered again individually. Finally, the clustering with lowest costs is reported.

4.2.1   Layout-based heuristic

The goal of this approach is to arrange the vertices in a n-dimensional space, such that the layout reflects the similarity values. Subsets of nodes with high edge-density should be arranged next to each other, and far away from other nodes. Afterwards the positions are used to cluster the data with geometrical clustering algorithms. TransClust currently integrates two layout methods: the first algorithm moves the nodes based on attractive and repulsive forces between the nodes and the second is inspired by ant colony behavior. While the force-based approach always produces the same layout for a given initial layout and a fixed set of parameters, the ant colony method depends on probabilities and hence leads to different solutions for each

4.2. CLUSTERING METHODS

run. TransClust is implemented such that it can easily be extended and improved. This means that practically any graph layout algorithm and geometric clustering algorithm can be integrated. Moreover, it is possible to combine different approaches. One algorithm may be the initial layout for another, in order to increase the performance of the subsequently used geometric clustering methods. In this manner, it is possible to create a list of different layout algorithms. This section describes the two implemented layout algorithms as well as the integrated geometric clustering methods. Since both layout methods have several parameters, an evolutionary parameter training has also been integrated into TransClust. Figure 4.4 depicts the workflow of the layout-based heuristic as it is implemented in TransClust.

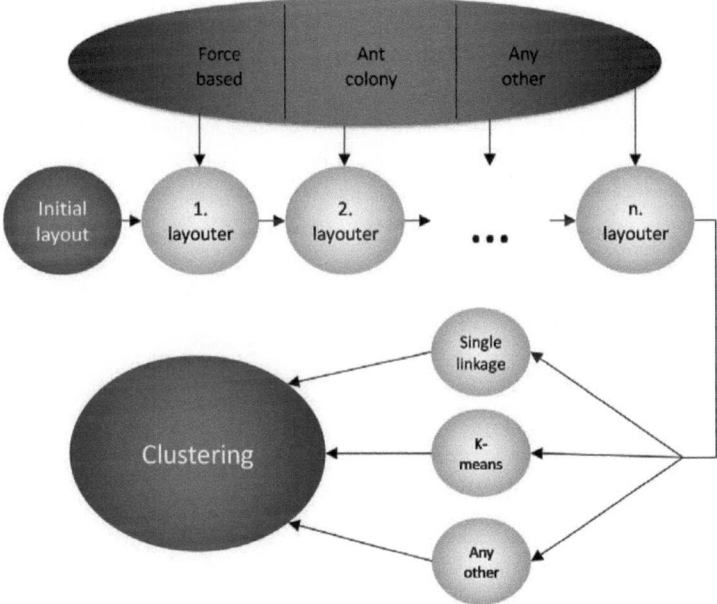

Fig. 4.4: Illustration of the layout-based heuristic as it is implemented in TransClust. Starting with an initial layout, a list of layout algorithms can be processed consecutively. The final layout is used to create the cluster assignments with geometric clustering methods.

Initial layout

Most graph layout algorithms depend on a starting point after which the position of each node is altered until a desired positioning is reached. Such an initial layout can be randomly chosen, but is preferably always the same for the same data to allow reproducibility of the results. The default initial layout in TransClust is to arrange the nodes in circles. For n dimensions $\binom{n}{2}$ circles are created, one for each pair of dimensions. The nodes are also split into $\binom{n}{2}$ equally sized groups respecting the order in which they were read as input. Each group is assigned to one circle and its elements are arranged on the circle, such that all neighboring nodes are equally distant from one another. Although the nodes are not as equally distributed over the hypersphere as on a circle in 2 dimension, this methods guarantees reproducibility and the initial layout can be computed quickly.

Force-based layout

The main idea of the force-based layout approach is to iteratively change the position of each node, such that nodes corresponding to similar objects move towards each other while nodes corresponding to dissimilar objects diverge. The implemented algorithm is hence a customized version of the force-based layout algorithm by Fruchterman and Reingold [35]. To reflect the underlying WTGPP, the nodes affect each other depending on their similarity, the threshold of the problem, and the current position of the nodes. For a user-defined number of iterations R, the interaction between every pair of nodes and thus the displacement for every node is calculated. All nodes are then simultaneously moved to their new position. Since this model is only inspired by physical forces without friction, it does not include acceleration. As a first step, before the layout method applies, the similarities are normalized in order to produce similarly good results for different applications using the same set of parameters. The displacements

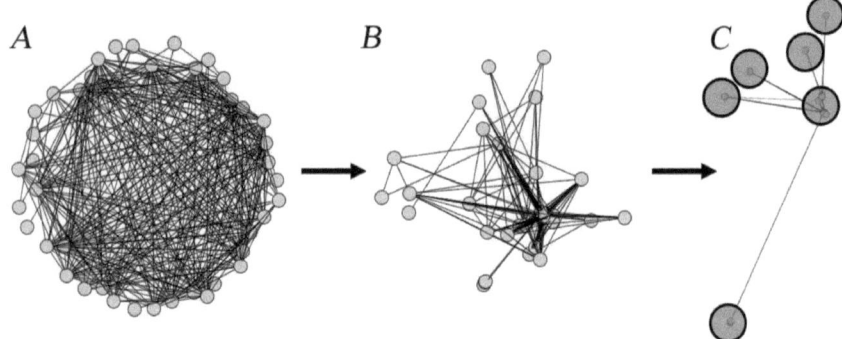

Fig. 4.5: An illustration of the force-based layout after (A) 10 iteration, (B) 30 iteration, and (C) 100 iterations. The circles in (C) represent putative clusters.

are computed according to the WTGPP, where nodes that are connected by an edge attract each other and those that are not adjacent repel each other. As described in Algorithm 1, the strength $f_{u \leftarrow v}$ of the effect of one node v to another node u (i.e., the magnitude of the displacement of u caused by v) depends on the Euclidean distance $d(u,v)$, on the cost to add or delete the edge and a user defined attraction or repulsion factor f_{att}, f_{rep}:

$$f_{u \leftarrow v} = \begin{cases} \dfrac{cost(uv) \cdot f_{att} \cdot \log(d(u,v)+1)}{|V|} & \text{for attraction,} \\ \dfrac{cost(uv) \cdot f_{rep}}{|V| \cdot \log(d(u,v)+1)} & \text{for repulsion.} \end{cases}$$

4.2. CLUSTERING METHODS

One can see that with increasing distance, attraction strength increases while repulsion strength decreases. This allows nodes which are connected by an edge to reach one another even if they were initially arranged far apart. Respectively, the repulsion effect is stronger for nearby nodes, so that the system is stable and the possibilities of both, adding and deleting edges, are treated equally.

In order to improve convergence to a stable position with minimal interactions, a cooling parameter is added. In practice, this means that the displacement is limited to a maximal magnitude M_i in each iteration i. M_i starts at an initial value M_0 and decreases with every iteration i:

$$M_{i+1} = \frac{M_0 \cdot |V|}{(i+1)^k}$$

k is a variable parameter, which is set to 2 by default based on empirical studies. Algorithm 1 shows the pseudo code for this algorithm, and Figure 4.5 provides an exemplary illustration of progress over time. TransClust allows, in contrast to its predecessor FORCE, multiple dimensions to avoid conflicting interaction between objects.

Algorithm 1 Graph layout

Input: similarity matrix $(S_{ij})_{1 \leq i < j \leq n}$ with $S_{ij} := \text{sim}(ij) - t$; circular layout radius ρ, attraction factor f_{att}, repulsion factor f_{rep}, number of iterations R
Output: node positions $pos = (pos[1], \ldots, pos[n])$; each $pos[i] \in \mathbb{R}^n$.
1: $pos = arrangeAllNodesCircular(\rho)$ ▷ initial layout
2: **for** $r = 1$ to R **do**
3:     ▷ Compute displacements Δ for iteration r
4:     initialize array $\Delta = (\Delta[1], \ldots, \Delta[n])$ of displacement vectors to $\Delta[i] = (0,0)$ for all i
5:     **for** $i = 1$ to n **do**
6:         **for** $j = 1$ to $i - 1$ **do**
7:             **if** $S_{i,j} > 0$ **then**
8:                 $f_{i \leftarrow j} = \log(d(i,j) + 1) \cdot S_{i,j} \cdot f_{\text{att}}$ ▷ attraction strength
9:             **else**
10:                 $f_{i \leftarrow j} = (1/\log(d(i,j) + 1)) \cdot S_{i,j} \cdot f_{\text{rep}}$ ▷ repulsion strength
11:             $\Delta[i] \mathrel{+}= f_{i \leftarrow j} \cdot (pos[j] - pos[i])/d(i,j)$
12:             $\Delta[j] \mathrel{-}= f_{i \leftarrow j} \cdot (pos[j] - pos[i])/d(i,j)$
13:     ▷ Move nodes by capped displacement vectors
14:     **for** $i = 1$ to n **do**
15:         $\Delta[i] = (\Delta[i]/\|\Delta[i]\|) \cdot \min\{\|\Delta[i]\|, M(r)\}$
16:         $pos[i] \mathrel{+}= \Delta[i]$
17: **return** pos

Ant colony layout

The second layout method is based on ant colony behavior. Virtual ants run in the layout space and can pick up an object and drop it at a different position depending on the neighborhood. Over time dissimilar nodes are moved away from each other and put together with similar nodes. In practice a grid is used for simplification. The virtual ant jumps to an arbitrary node and decides whether to pick it up or not, depending on all nodes in its neighborhood. One parameter of this heuristic is the size of the neighborhood, which has an impact on quality and running time. The decision about changing the position of a node u is modeled according to probabilities. It is more likely for a node to be picked up if most nodes in its surrounding area are not similar to it. By contrast, the probability is low if similar objects can be found in the neighborhood. The pick-up probability is defined as:

$$p_{\text{pick}}(u) = \left(\frac{k^+}{k^+ + f(u)}\right)^2$$

where k^+ is a user defined parameter and f is a function to judge the neighborhood of u defined as:

$$f(u) = \max\left(0, \frac{1}{\sigma^2} \sum_{v \in N}\left(1 - \frac{1 - c'(u,v)}{\alpha}\right)\right)$$

where N denotes the set of items in the local neighborhood and σ^2 its size. $c'(u,v)$ are the costs for adding/deleting the corresponding edge, normalized to $[-1, 1]$ by dividing through the maximal costs that appear in the graph. α is a variable parameter to control the influence of the costs on the probability.

Carrying around this object, the ant randomly walks around and drops the object after each step with a drop probability p_{drop} again depending on the other nodes within the area around the ant:

$$p_{\text{drop}}(u) = \left(\frac{f(u)}{k^- + f(u)}\right)$$

where f is defined as above and k^- is another user-defined parameter for this approach.

A detailed description of this layout algorithm can be found in [50]. This heuristic may produce different results for each run, because it integrates randomized parts. The ant colony layout has been initially developed as an alternative to the force-based layout method or as a pre-processing step. Unfortunately it was shown that this layout approach performs worse than the force-based layout and could also not improve the performance of the force-based method when used as initial layout. However, it demonstrates the capability of TransClust to support various layout methods and the combination of them.

Geometric single linkage clustering

Given a layout of the nodes in the layout space $\mathbb{R}^m$, geometric single linkage clustering is used to assign the nodes to clusters.

For a fixed radius r the nodes are assigned as follows: Starting with an arbitrary node n_1, all nodes with distance smaller than r are assigned to the same cluster. Subsequently these nodes are used as new seeds. That is, all nodes that have distance smaller than r to the new seeds are also assigned to the same cluster, and so on, until no new nodes within range can be found. Again, from all remaining nodes, which have no cluster assignment yet, a new seed is chosen randomly and the previously described steps are repeated until every node is assigned to exactly one cluster.

4.2. CLUSTERING METHODS

To find the best clustering according to the objective function multiple radii are used. A sorted list of all distances between nodes in the graph is partitioned into a pre-defined number of k parts. The cost for an assignment using the k center distances of each part are calculated. Recursively the part with the best cost is partitioned again into k parts. This procedure is not guaranteed to find the best distance, but is efficient and returns good solutions in practice. For a graph with n nodes, at most $k \cdot \text{argmax}_{i \in \mathbb{N}} \{k^i < n\}$ complete assignments and subsequent calculation of costs are necessary.

K-means clustering

An alternative geometric clustering approach is k-means. It was integrated into TransClust as an alternative to the geometric single linkage clustering method. Here, a customized version of the standard k-means approach is used. The starting points are not chosen completely random, but are positions of random nodes. Since the number of clusters that optimize the objective function is not known in advance the clusters are calculated for different k, ranging from 1 to $\max\{n, k_{\max}\}$, where n is the number of nodes in the cluster and $k_{\max}$ is a pre-defined maximum. It is necessary to restrict the maximal number of clusters ($k_{\max}$), due to the long runtime that would otherwise result for large problem instances. However, this restriction can be compensated with the post-processing, particularly with recursive reclustering (see Section 4.2.4 for a description of that post-processing method).

In [46] a comparison between the two geometric clustering approaches showed only minor differences in the runtime and quality. This indicates, that the choice of the geometric clustering method is not as important as the choice of a good layout algorithm. However, it is possible to integrate any geometric clustering method into the TransClust framework. The default method in TransClust is the geometric single linkage clustering.

Parameter optimization

There are several parameters influencing the runtime and quality of a heuristic. The force-based layout, for instance, needs the number of iterations R, the attraction and repulsion scaling factors $f_{\text{att}}, f_{\text{rep}}$, and the magnitude M_0 to be specified. A practical method for finding problem specific values is evolutionary training. TransClust implements such a strategy in two different ways.

First, a good parameter combination is determined that can be applied to most of the graphs. This is done by a pre-computation on a training data set. Since, however, the optimal parameter constellation depends on the specific problem, one can apply such a training algorithm to each individual problem (connected component). TransClust allows the specification of the number of generations to train, and thus to adjust runtime and the quality of the result.

Training works as follows: First, one starts with a set of 15 randomly generated parameter sets and the initial parameters mentioned above. The parameter sets are sorted by the cost of solving the WTGPP on the given graph. For each generation, the best 10 parameter constellations are used as parents to generate 15 new combinations. In order to obtain fast convergence into a good constellation, as well as a wide spectrum of different solutions without running into local minimum, TransClust splits these 15 new combinations into 3 groups with 5 members each. The first group consists of parameters obtained only by random combinations of the 10 best already known parameter constellations. The next group is generated with random parameters, while the third group is obtained by a combination of the previous methods. To reduce the runtime for computing solutions of problems that are small or easy to compute, a second terminating condition is added: If at most two different costs appear while calculating the 15 start parameters, the best one is chosen, in which case no more generations are computed.

4.2.2 Integration of the CAST algorithm

The CAST algorithm is a quick heuristic for finding a solution of a WTGPP. A slightly changed implementation of this algorithm is integrated into TransClust for a quick approximation of an upper bound of costs necessary to make the input graph transitive. In contrast to the original method, the choice of the first element in one cluster is not selected randomly. Instead, all elements are sorted in descending order by their average similarity to all other elements. The first element of each cluster, is the first element of the sorted list that is not yet assigned to a cluster. For a description of the CAST implementation in TransClust see Algorithm 2. The TransClust environment benefits from the computed boundary in two ways: First, data reduction rules as used in the FP algorithm can be applied. Hence, those nodes for which the removal of their connecting edge would cost more than the costs computed with CAST, are merged. It has been shown that these reduction rules are most efficient for medium size graphs (refer to [24]). Consequently they are applied in TransClust on instances with up to 200 objects. Second, the runtime of the FP algorithms depends on a good guess for an upper limit. The closer the approximation is to the real costs, the fewer unnecessary branches must be made. In practice, a fraction of the costs calculated by CAST is taken as a starting point for the FP approach and increased until a solution is found.

Due to its different nature, the CAST heuristic also sometimes reveals better results than the layout-based heuristic. Thus, both approximation algorithms are executed and the best solution is reported (refer to Section 4.4.3 for a comparison of the heuristic methods in TransClust).

4.2.3 Integration of the exact fixed parameter approach

As previously described in Section 3.4.1 an exact branch and bound strategy for solving the WTGPP was developed by Böcker *et al.* at Jena University. It will be shown in later evaluations (see Section 4.4) that the fixed-parameter approach is applicable for small instances of the problem. The runtime strongly depends on the specific graph and not only on its size, making it often unpredictable how much computation time is needed for one instance.

TransClust takes advantage of this approach by integrating it and restricting its usage to a maximal instance size and a maximal time for each calculation. A simplified version of the algorithm is implemented in Java, designed to work with the data structures that are present in the TransClust framework.

4.2. CLUSTERING METHODS

Algorithm 2 Implementation of the CAST algorithm in the TransClust framework
Input: similarity matrix $(S_{ij})_{1 \leq i < j \leq n}$ with $S_{ij} := \text{sim}(ij) - t$
Output: clustering assignment for each object $c = (c[1], \ldots, c[n])$; each $c[i] \in \mathbb{N}$.
1: initialize c with $c[i] = -1$ for all $i = 1, \ldots, n$
2: initialize list $remaining$
3: add $i = 1, \ldots, n$ to $remaining$
4: sort $remaining$ in descending order by their average similarity to all other nodes
5: initialize integer $currentClusterNumber = 0$
6: **while** $remaining$ is not empty **do**
7:    **for** $i = 1$ to n **do**
8:       **if** $i \in remaining$ **then**
9:          $c[i] = currentClusterNumber$
10:         remove i from $remaining$
11:         $currentClusterNumber = currentClusterNumber + 1$
12:         initialize double $costChange = 0$
13:         Repeat ADD and REMOVE as long as changes occur
14:         ▷ ADD: add all elements that reduce costs by adding to cluster of i
15:         **while** $costChange \leq 0$ **do**
16:            integer $best = \text{argmin}_{j \in remaining} \left(\sum_{\{k; c[k] = c[i]\}} -S_{kj} \right)$
17:            double $costChange = \min_{j \in remaining} \left(\sum_{\{k; c[k] = c[i]\}} -S_{kj} \right)$
18:            **if** $costChange < 0$ **then**
19:               remove $best$ from $remaining$
20:               $c[best] = c[i]$
21:            $costChange = 0$
22:         ▷ REMOVE: remove the worst elements from the cluster of i if this reduces the costs
23:         **while** $costChange \leq 0$ **do**
24:            integer $best = \text{argmin}_{j; c[j] = c[i]} \left(\sum_{\{k; c[k] = c[i]\}} S_{kj} \right)$
25:            double $costChange = \min_{j; c[j] = c[i]} \left(\sum_{\{k; c[k] = c[i], k \neq j\}} S_{kj} \right)$
26:            **if** $costChange < 0$ **then**
27:               add $best$ to $remaining$
28:               $c[best] = -1$
29: **return** c

Each instance of a smaller size than a pre-defined threshold is processed using the FP approach. If the exact algorithm does not find a solution in a user-defined maximal time it is aborted and the heuristic methods are applied instead. It is necessary to restrict the usage of the fixed-parameter approach in order to produce results in reasonable time.

The integration of this exact method is particularly beneficial in the recursive post-processing method, which is described in the next section. The problem instances, which have to be solved there, are usually much smaller in size and have a structure which can be solved efficiently with the fixed-parameter method.

4.2.4 Post-processing

Small variations in a clustering can increase the value of the objective function enormously. If, for instance, one node is assigned to a "wrong" cluster, all edges between that node and its cluster, as well as that node and the "better" cluster, are affected. To avoid such small mistakes, different post-processing method have been implemented into the framework, which improve the output of a given heuristic. All three of these integrated methods (merge, rearrange single nodes, and recluster recursively) are described below. However, TransClust uses a single method that combines all three of the post-processing strategies. This method produces good results in almost the same time as the individual strategies.

In the following let $C = \{C_1, \ldots, C_n\}$ be the clustering used as input for every post-processing method.

Merge The merge strategy iterates through every pair of clusters $C_i, C_j \in \binom{C}{2}$ and calculates the costs of the clustering $C_{i,j} := C \cup \{C_i \cup C_j\} \setminus \{C_i, C_j\}$. If there exist two clusters C_i and C_j, such that the costs for producing the clustering $C_{i,j}$ is smaller than the costs for C, the pair which produces lowest costs is merged. The resulting clustering is used as input for an additional iteration until no merging any pair would lower the overall costs. Note that it is not necessary to compute the complete costs, but only the change of costs, i.e. the costs to add all edges between the two clusters of interest.

Rearrange single nodes To further improve the results of a clustering obtained from a heuristic, this method tries to change the assignment for every single node. For one iteration it calculates for each node v and each cluster C_i the costs for moving v to C_i. The best movement is performed, if it reduces the overall costs. As in the previous method, the modified graph is again input for an additional iteration of this method until no further changes provide a reduction of costs. Note that a node can also be assigned to an empty cluster and thus create a new singleton and a singleton can be removed by assigning the corresponding node to a different cluster. In the current implementation this method and the previously described merging alternate until no changes are made in either of the two methods.

Recluster recursively The last post-processing method takes advantage of the fact that the heuristics perform better on smaller instances and the exact solution may be calculated for small problems with the integrated fixed-parameter approach. For every cluster C_i a similarity graph $G_i = (C_i, E_i) \subset G$ is defined as the induced subgraph of the initial graph G:

$$G_i = (C_i, E_i); \quad E_i = \left\{u, v \in \binom{C_i}{2}; \text{sim}(uv) > t\right\}$$

Depending on the sizes of G_i either a heuristic or the exact approach are used to solve the corresponding WTGPP of the subgraph. This step can be repeated recursively for the results for each G_i and stops if the subgraph cannot be split any further. Figure 4.6 illustrates all of the post-processing steps as they are implemented in TransClust.

4.2. CLUSTERING METHODS

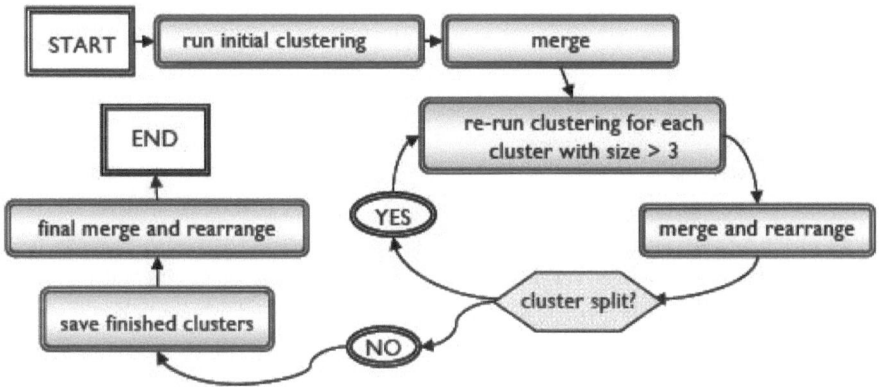

Fig. 4.6: An overview of the recursive post-processing method. Taken from [46]

4.2.5 Extensions and integration of existing knowledge

TransClust integrates the extensions to TC as described in the previous chapter. This includes the alternative clustering strategies:

- Top-down hierarchical clustering
- Bottom-up hierarchical clustering
- Overlapping clustering with fuzzy associations
- Overlapping clustering by single rearrangement

Furthermore it is possible with TransClust, to integrate existing knowledge into the clustering process to improve the clustering quality. This can be done in the following ways:

- Define upper/lower bound
- Set known assignments
- Set/forbid any pair of objects to be in one cluster

4.2.6 Threshold determination and supporting analyses

Finding the right density parameter for a clustering problem is challenging and problem specific. TransClust supports the detection of a meaningful density parameter in different ways. On the one hand it is possible with TransClust to cluster a given input data with several thresholds. The distribution of each run is presented and aids specifying the best threshold, if knowledge about the expected sizes of the clusters is known. A second option is to use existing knowledge about parts of the data or about a related problem. Again the data can be clustered for a set of thresholds but in addition it is compared to the known assignments. The F-measure is used as quality measure and reflects how close a clustering with a certain threshold is to the gold standard clustering. Furthermore, the Cytoscape plugins of TransClust allow for investigation of the similarity function and also help in finding the right threshold. Histograms illustrate the distribution of similarities within a cluster and between two clusters. Since the density parameter of TC is a similarity threshold, the interesting range, where the best threshold can be found, is detectable using these plots.

Figure 5.5 on page 80 illustrates such a typical analysis for a given problem using the TransClust plugin for the network visualization and analysis framework Cytoscape, exemplarily for the task of sequence clustering. For this example it is assumed that a gold standard assignment is known for a subset of the input. The corresponding analysis can be found in Section 5.2.3.

4.3 Availability

Two important focuses during the development of TransClust were its availability and applicability. On the one hand this means that the software should be free and available online without requiring the purchase of any additional software. On the other hand it means that the software must be intuitive to use. TransClust is implemented in Java to allow for execution on any operation system. TransClust can be accessed in three different ways: as a standalone version, which can be executed via the commandline or with a graphical user interface, as an integration into the network analysis and visualization software Cytoscape, and as a web application. This section introduces to these three options. For large problem instances it is recommended to use the standalone implementation of TransClust, since this option is optimized for memory and runtime efficiency.

4.3.1 Standalone application

The standalone implementation of TransClust provides two ways for running the clustering procedure. One can either start TransClust from the commandline or use the implemented Graphical User Interface (GUI). This section describes how to perform a clustering, using the GUI.

The graphical user interface of TransClust is organized in three panels; (1) the parameter panel, (2) the preview panel, and (3) the console panel (see Figure 4.7):

Parameter panel This panel contains three tabs to specify the different parameters of each method.

- The first tab is the clustering tab. From here one can cluster data from a similarity file with a list of threshold. The range of thresholds and the step size between different thresholds has to be specified here as well. Furthermore, one can decide which kind of clustering should be performed: a partitional clustering solving a WTGPP for each threshold, an hierarchical clustering using the bottom-up or top-down approach, or one of the two overlapping methods presented in Section 3.3.3.

4.3. AVAILABILITY

An upper and lower bound can be set in this tab as well, to influence the clustering results. To start a clustering process from this tab a similarity file has to be loaded. Either it has been directly specified through the menu or calculated using the various similarity functions for sequence similarity on a BLAST and a FASTA file.

- The second tab is the clustering-parameter tab, where the parameters of the heuristics and exact algorithm can be specified. One can choose to use only the faster but less accurate CAST approximation, or the CAST and the more precise layout-based heuristic. The time and maximal size of a component for applying the exact fixed-parameter algorithm can be specified in this tab as well. Further parameters are the choice of the layout methods, the geometric clustering method of the resulting layout, the number of dimensions of the layout algorithm, whether post-processing should be applied and which method, and the number of CPUs that should be used for a parallel computation of the connected components. The default parameters are set to be a good trade-off between speed and quality and should be used mostly. Note that starting the clustering process from this tab requires to specify a directory with costmatrices or to import/calculate costmatrices from a similarity file or from a BLAST and FASTA file. All specified clustering parameters will be used in any clustering process, including those started from a different tab.

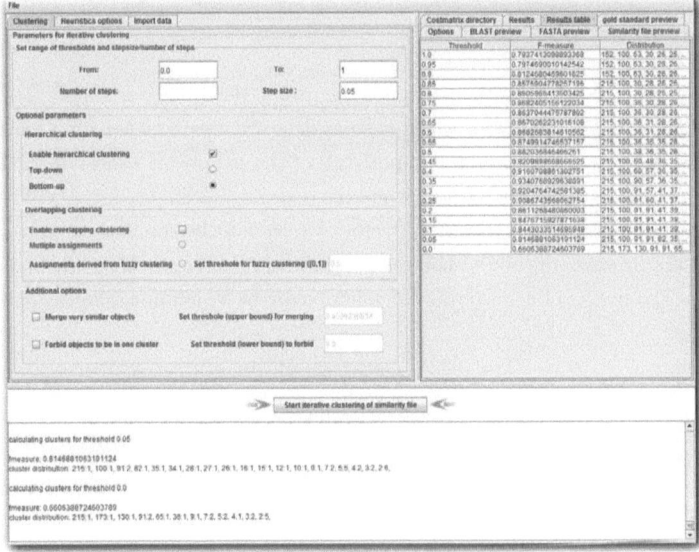

Fig. 4.7: The graphical user interface of TransClust. Three panels are shown; (1) the parameter panel (left), (2) the preview panel (right), and (3) the console panel (bottom). Files can be imported and saved through the menu (top).

- The last tab in this panel is the import tab. After a tab delimited similarity file, or a

BLAST and FASTA file are loaded via the menu this tab offers to specify parameters for creating costmatrices for a subsequent clustering. For similarity files simply the threshold and an upper bound as described above can be chosen. If a BLAST and FASTA file are loaded one has to choose between the different similarity functions and has to specify an E-value cutoff. Every High Scoring Pair (HSP) with higher E-value than this cutoff will be ignored in the import. This might be especially interesting if the score function is chosen (see Section 2.2.2 for more details about the different similarity functions).

Preview panel Like in the previous panel the preview panel is also organized in tabs. As soon as any file is loaded a preview of the first 100 rows can be seen here. Furthermore the results of a clustering or an iterative clustering for different thresholds will be displayed in this panel as a preview of the result file and a sortable table. One can check if the loaded files have the correct format, and which files are loaded so far. The panel automatically updates if any changes are made or a clustering process is finished.

Console panel The console offers information about the current process. If a clustering process using costmatrices is started, the progress and clustering informations like time, score, size of connected component and cluster distribution of the result are displayed. For an iterative or hierarchical clustering, only the current threshold, the cluster size distribution for this threshold, and the calculated F-measure are shown.

The last item to describe in this section is the menu. Here, one can import similarity files or BLAST and FASTA files, from which costmatrices are created. A popup window opens and provides the same options as the import tab of the parameter panel. If a BLAST and FASTA file are imported the similarity file and the costmatrices are produced. All files created in TransClust are stored in a temporary directory, which must be specified on startup. To load an already existing similarity file without creating costmatrices, one can use the load option in "File" menu. The results of a clustering can additionally be stored in a different location.

4.3.2 Cytoscape plugin

Cytoscape is a network analysis and visualization platform that assists in the analysis of a wide spectrum of network data [29]. Multiple formats for importing data are supported and various layout options for a network are provided. Furthermore, it allows developers to implement plugins for specific tasks.

The main advantage that Cytoscape provides for a cluster analysis is the visualization of the network. To use this feature together with TC, TransClust has been implemented as a plugin for Cytoscape. Furthermore two additional plugins have also been developed: the BLAST2SimGraph plugin and the ClusterExplorer plugin. Cytoscape supports various formats for importing networks from tables or other formats, but does not include a similarity calculation between pairs of protein sequences. This gap was closed with the BLAST2SimGraph

4.3. AVAILABILITY

plugin to allow a cluster analysis of sequences from within Cytoscape. It reads an all-vs.-all BLAST result file and the corresponding sequences in FASTA format and calculates a pairwise similarity based on the function described in Section 2.2.2. The ClusterExplorer plugin offers various methods for analyzing a given clustering. Most important is a visualization of the intra vs. inter edge distribution, which, applied on a gold standard clustering, can aid with the identification of a reasonable threshold and may also be used to evaluate whether a similarity function can be used for a given task or not. Additional methods within the ClusterExplorer plugin are the comparison between two clusterings, identifying the center of each cluster, and calculating the distance between two clusters or between a object and a cluster. Furthermore the cluster size distribution can be visualized as a histogram as well as the overall edge weight distribution.

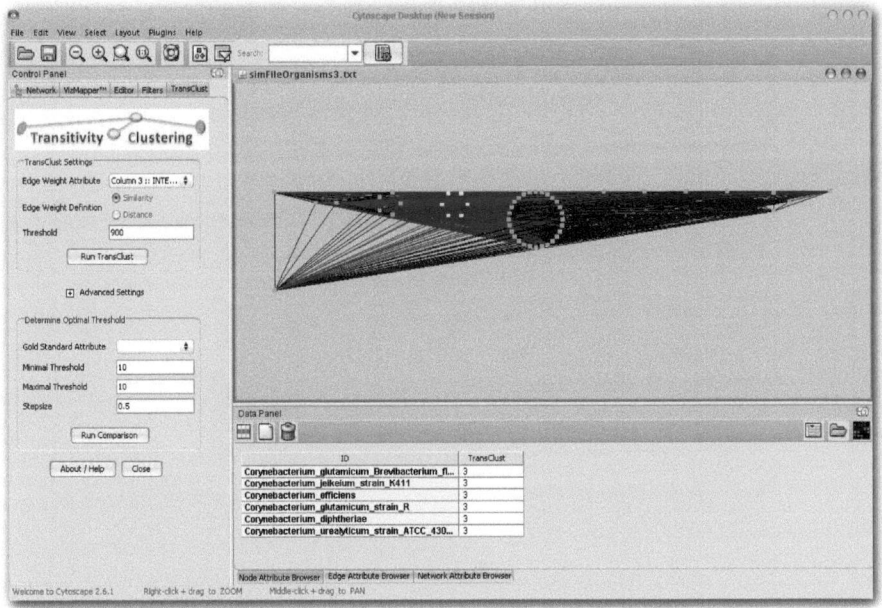

Fig. 4.8: Illustration of the Cytoscape software with integrated TransClust plugin (left side). The right side shows a visualization of a similarity graph that was clustered with the TransClust plugin.

The TransClust plugin has similar features to the standalone version of TransClust. One can cluster a network with a specific threshold or use multiple threshold and compare the clustering results against a gold standard. A typical analysis of protein sequences using these plugins is shown in Figure 5.5 on page 80. Note that Cytoscape needs a lot of memory space and is thus not applicable for large scale analysis of data. However, for a medium sized problem the additional visualization and the ClusterExplorer features are very useful. Cytoscape is available at http://www.cytoscape.org and the aforementioned plugins can be downloaded from http:

//transclust.cebitec.uni-bielefeld.de. This site also includes an online tutorial, which guides through the usage of TransClust and the Cytoscape plugins.

4.3.3 Web application

TransClust is available as a web application at the TransClust website. For problem instances of up to 400 nodes this application provides an easy-to-use method to quickly identify clusters. To use the web application one can upload a tab delimited similarity file, a BLAST and a FASTA file to cluster sequence data, or a costmatrix file. Due to server restrictions, it is only possible to cluster small sets of objects. For larger applications it is recommended to either use the standalone version of TransClust or download Cytoscape and the TransClust plugins for a visual analysis of the data. The web page directly links to a Java web start application, that starts the standalone application with graphical user interface as applet.

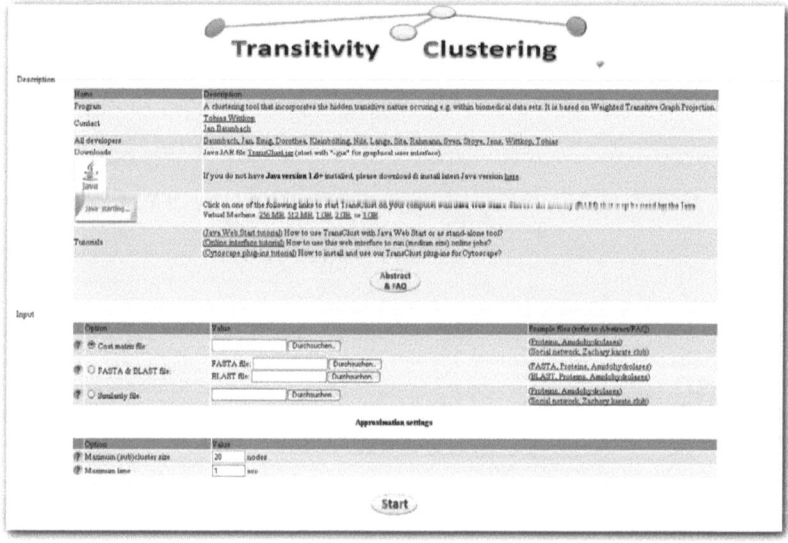

Fig. 4.9: Screenshot of the TransClust website. From this site one can start the web application (bottom). For larger files it is recommended to use the standalone version, which can be executed from this site as Java web start and hence runs on the client side. A tutorial explains how to use TransClust, including some example applications. The source code of TransClust and the Cytoscape plugins are available here as well.

4.4 Evaluation of the integrated TransClust framework

This section includes evaluations of the TransClust framework. First the different methods in TransClust are compared against each other to explain the default configurations. Furthermore the influence of post-processing on the two heuristic methods of TransClust is analyzed. This

4.4. EVALUATION OF THE INTEGRATED TRANSCLUST FRAMEWORK

section concludes with a performance evaluation of TransClust in comparison to the exact FP algorithm developed by Böcker *et al.*

4.4.1 Data

For this study real world data from protein sequence comparisons as well as artificial data has been used.

Real world data obtained from COG

To have a sufficiently large set of connected components, and thus instances of the WTGPP, protein sequences from all 66 prokaryotic genomes of the COG database (2007) have been used for this study. BLAST [2] was performed on the corresponding sequences with an E-value cutoff of 0.01. Subsequently the pairwise similarities are calculated using the $-\log$ of the E-values as described in Section 2.2.2 (BeH). A similarity threshold of 10 produces problem instances with sizes varying from singleton clusters to instances with several thousand nodes. Table 4.1 summarizes the used dataset.

Dataset	All proteins in COG (2007)
Similarity	BeH (best $-log$(E-value))
Threshold	10
Intransitive connected components	4,019
No. of proteins in transitive components	55,185 (31.3 %)
Size of largest component	4,410
Total no. of proteins	176,104

Tab. 4.1: The dataset used for runtime and quality analysis of TransClust heuristics against fixed-parameter method. Taken from [46]

Artificial data

As a second data set, random artificial graphs were created. Given the number of nodes n, an integer $k \in [1, n]$ is randomly selected (uniform) and hence defines a cluster of the first k nodes. The remaining $n - k$ nodes are processed as above until no nodes are left. This gives a random number of clusters of random sizes. The similarities of objects within a cluster are then drawn from a Gaussian distribution $\mathcal{N}(\mu_{in}, \sigma_{in}^2)$; they are positive on average, but negative with some probability. Similarities of objects in different clusters are conversely drawn from a Gaussian distribution $\mathcal{N}(\mu_{ex}, \sigma_{ex}^2)$, which leads to negative values on average. If the parameters are chosen carefully, this construction leads to almost transitive graphs. For the evaluation in this section the parameters are $\mu_{in} = 21$, $\mu_{ex} = -21$, and $\sigma_{in} = \sigma_{ex} = 20$, so that the probability of seeing an undesired or missing edge is about 0.147 per node pair. With these parameters ten graphs for each of the sizes 10 to 300 in steps of 10 were constructed.

maximal size	maximal time	costs	time
0	0	4275795.43	20min 26s
20	1	4275691	20min 34s
50	1	4275295.13	20min 55s
100	1	4275263	26min 49s
100	5	4275195.00	33min 33s
200	10	4273090.19	1h 1min 31s

Tab. 4.2: Costs and time for clustering of the COG data set using different limitations for the exact FP approach

4.4.2 Optimizing the combination of methods in TransClust

TransClust optimizes the trade-off between accuracy and running time. This means on the one hand, that the heuristic methods themselves have to be optimized and on the other hand, that the decision when to use which approach has to be specified.

The CAST heuristic does not need any specification of parameters. In the layout-based approach one has to decide which layout method and which geometric clustering should be used.

In [50] a comparison between the ant colony layout (ACL) and the force-based layout has been performed. This evaluation demonstrated that the force-based approach clearly outperforms ACL. For this reason, and to avoid irreproducibility, due to the random aspects of ACL, the force-based layout is the default layout method in TransClust. The parameters of this method, namely the attraction factor f_{att}, the repulsion factor f_{rep}, the magnitude M_0, and the number of iterations R have been optimized by using the integrated parameter training on a subset of the COG data set. The resulting parameters are: $f_{att} = f_{rep} = 16$, $M_0 = 100$, and $R = 80$. The layout dimension is set to 3, based on a study performed in [46].

The two different geometric clustering approaches K-means and single linkage clustering were already compared in [46]. Both approaches have a similar runtime, but single linkage clustering reveals better results and does not include any random mechanism. It is thus the default method in TransClust. The results of the aforementioned comparisons can be found in the appendix.

The last decision that has to be made is about the limitations of the exact FP algorithm. Table 4.2 summarizes the results of a corresponding test using different limitations. A good trade-off between time and quality can be achieved with a maximal size of 50 nodes per instance and a maximal time of 1 second. For small instances, the FP approach is even faster than the heuristic methods. Although the overall costs may change only slightly, these changes occur mostly in small graphs, where changes do not infect the costs as much as in large graphs. If a higher accuracy is wanted it is still possible to change the settings. It is recommended in such cases to use 200 nodes as maximal size and 10 seconds as maximal time.

4.4. EVALUATION OF THE INTEGRATED TRANSCLUST FRAMEWORK 67

4.4.3 Influence of post-processing on accuracy

In order to see the impact of the post-processing on the quality of the results of the heuristic methods implemented in TransClust, an evaluation with the previously described COG data set has been performed. The methods used in this evaluation are the force-based heuristic and the CAST algorithm. As the post-processing method, the recursive post processing method is used to compare the differences in the resulting costs.

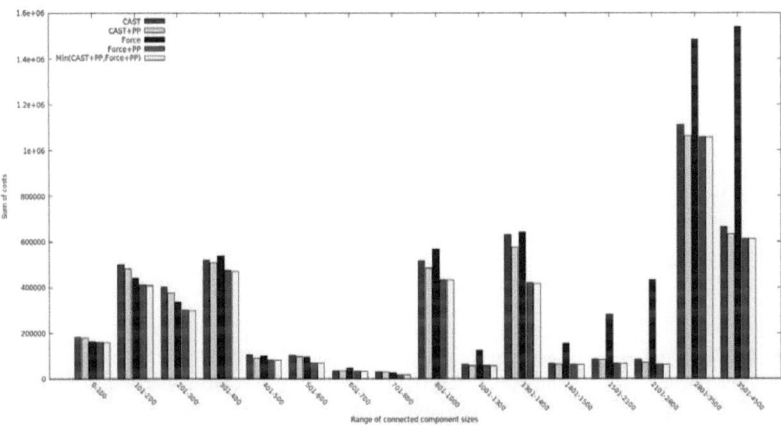

Fig. 4.10: Comparison of the CAST heuristic and the force-based heuristic (force), using either no post-processing (PP) or the default recursive post-processing method.

Figure 4.10 illustrates the results of this comparison for the COG data set. One can see that the post-processing improves the results of the force-based heuristic much more than those of the CAST algorithm. This might be explained by the methodical similarity between the post-processing method and the CAST approximation. Even if in some cases CAST initially produces results with lower costs than the layout-based algorithm, the best results can mostly be achieved by using the combination of the layout-based approach and the post-processing. However, it is not surprising that the two methods perform differently on different instances. This further confirms that it makes sense to combine different heuristic approaches to achieve the best possible results. While this section concentrated only on the comparison of the two main heuristic approaches integrated into the TransClust software, the next section evaluates the combined performance of these approaches in comparison to exact solutions.

4.4.4 Comparison against exact solution

This section presents an quality and runtime evaluation of the TransClust framework. For this comparison the current version of the FP branch and bound strategy is used and compared to the TransClust method. Note that is not necessary to include FORCE in this evaluation, since TransClust improved the used methods and adds new heuristics (see Figure D.1 and Figure

4. THE TRANSITIVITY CLUSTERING FRAMEWORK TRANSCLUST

D.2 for a comparison of FORCE and TransClust, performed in 2008). Furthermore, TransClust has been implemented from scratch and hence the used data structures where optimized for memory and runtime efficiency. In [65] it has been shown that FORCE outperforms the greedy approach by Marcel Martin (see Figure D.5 and Figure D.4 in the appendix for a runtime and quality comparison of FORCE, the greedy algorithm, and the fixed-parameter approach) and consequently this method is also excluded from the here presented evaluation. Although the Integer Linear Programming (ILP) formulation may be an alternative for calculating the exact costs for the WTGPP only the exact fixed-parameter approach is taken into account. This is done due to the similarity in maximal size and runtime of computable instances between the fixed-parameter and the integer linear programming methods (See Table 4.3 for comparison), and the fact that the ILP implementation needs the commercial CPLEX solver.

Size red. instances	3-49	50-99	100-149	150-199	200-249	250-299	300-1400
No. red. instances	297	52	16	10	9	2	19
Unfinished FP	0	0	1	1	2	2	15
time FP	125 ms	23.9s	44.1 min	4.52 min	47.3 min	n/a	8.98 min
Unfinished ILP	0	0	0	0	1	1	10
time ILP	17 ms	6.97 s	5.3 min	18.2 min	76.2 min	6.85 min	1.67 h

Tab. 4.3: Running times on protein similarity data after data reduction for fixed-parameter and integer linear programming approach. Running time for instances that did not finish after 24h were ignored for average running time computation. Taken from [24]

This test uses the COG data set and the artificially created data. Besides a quality test, the runtime of the used methods are compared to illustrate the necessity of using heuristic approaches for this problem. All tests are performed on the same machine to guarantee a fair analysis. Quality results are displayed for only a subset of the COG data: for some instances the exact solution could be found in reasonable time. The maximum time for solving a connected component was set to 2 hours, which excluded 26 graphs with sizes ranging from 140 to 4410 from the evaluation (26,471 elements in total sum), where FP could not find a solution in reasonable time. Note that even with a maximal time of 24 hours, the FP algorithm would not find solutions for most of these problems (see also Table 4.3).

Figure 4.11 summarizes the results of the quality analysis. Although there are some instances which were not solved correctly by the heuristic approach, the overall quality performance is remarkable. Out of almost 4000 connected components only 45 were not solved correctly with TransClust and even fewer (4) have a score difference of more than 10% to the exact solution. Note that some differences may occur, due to rounding errors. TransClust produces results with an overall difference of the score of less than 0.65% in a fraction of the time that would be required by the exact algorithm. When comparing the runtime of the methods used, it can be seen that it is neccesary to use heuristic approaches for problems of a larger size. Although a similarity graph can be split into connected components, and it is sufficient to solve these, the COG dataset shows that connected components with hundreds or even thousands of nodes occur in real-world applications. Nevertheless, it is possible to make use of a fast exact algorithm by

4.4. EVALUATION OF THE INTEGRATED TRANSCLUST FRAMEWORK 69

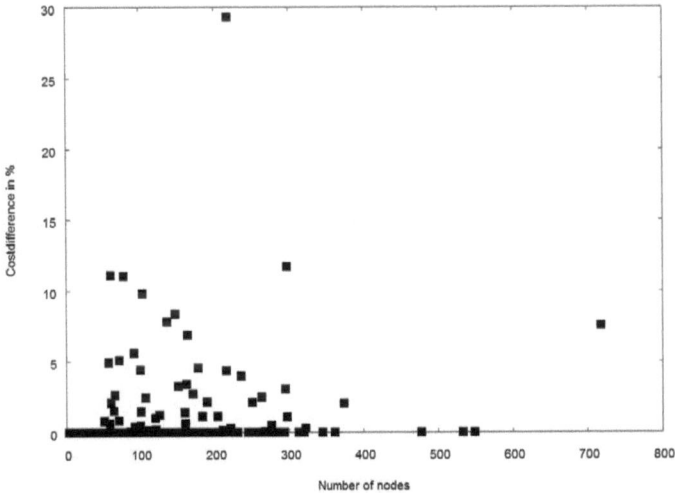

Fig. 4.11: Quality evaluation of the heuristic. On the x-axis the number of nodes is shown, while the y-axis is the distance to the optimal solution obtained with the fixed-parameter algorithm

combining it with heuristic methods. The performance of the FP approach can be significantly improved by guessing an upper limit close to the exact solution. In [24] it has been shown, that applying data reduction rules drastically reduces the complexity of most problems. Some of these rules rely on an upper limit as well and would benefit even more from a well estimated value. Still an exact algorithm cannot produce results for large instances in a reasonable amount of time even with such improvements. It is however possible to apply it to smaller instances. TransClust takes advantage of this fact by integrating the FP approach and limiting its usage.

Results for the artificial data set look similar to those of the real world data. Again TransClust clearly outperforms the exact algorithm in terms of runtime. The corresponding comparison can be seen in Figure 4.13. Note that it was not necessary to display a quality comparison, since TransClust was able to solve all instances correctly.

The overall finding of the evaluation of this section is that TransClust meets its objective of balancing quality with runtime, matching exact algorithms in accuracy and far outperforming them in speed.

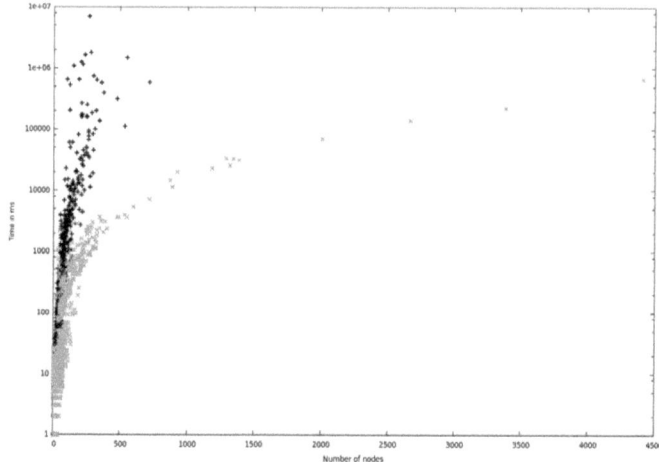

Fig. 4.12: Runtime comparison between the exact fixed-parameter algorithm (red) and TransClust (green) using the COG data set. The x-axis shows the number of nodes, while the y-axis shows the time in ms. Note that the y-axis is log scaled.

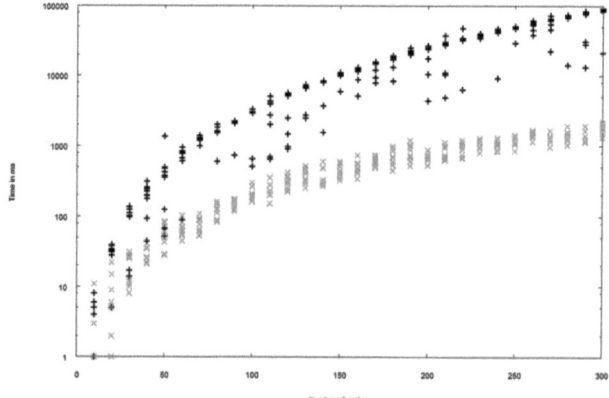

Fig. 4.13: Runtime comparison between the exact fixed-parameter algorithm (red) and TransClust (green) on the artificial data set. On the x-axis the number of nodes is shown, while the y-axis shows the time in ms. Note that the y-axis is log scaled

5. EVALUATIONS OF THE TRANSITIVITY CLUSTERING MODEL

As demonstrated in the previous chapter, TransClust is an efficient integrated method to accurately solve the Weighted Transitive Graph Projection Problem (WTGPP). The following sections are dedicated to real world evaluations.

The flexibility of Weighted Transitive Graph Projection (WTGP) allows it to be used for various applications. Throughout this section experiments on clustering protein sequences or predicting complexes in Protein-Protein Interaction (PPI) networks will be described. Transitivity Clustering (TC) is not the first clustering approach to be used for these tasks and, consequently, comparisons against commonly used approaches are necessary. It will be shown that TC can compete with the other clustering approaches and outperforms them in terms of quality. To ensure a fair comparison, previously performed evaluations are taken into account and TC is applied on the corresponding data sets using the corresponding quality measures. One such study is that of Brohée et al. [21], an evaluation of four clustering algorithms, applied to predict protein complexes in PPI networks. The clustering methods used in this study were Markov Clustering (MCL) [32, 76], Restricted Neighborhood Search Clustering (RNSC) [49], Super Paramagnetic Clustering (SPC) [19], and Molecular Complex Detection (MCODE) [7]. A critical point in such evaluations is the choice of the quality function. The chosen clustering problems have a reference assignment serving as the gold standard. Hence, external quality measures can be applied, meaning the clusters obtained from an algorithm are compared to the ground truth assignments. If different quality measures like recall and precision, or Positive Predictive Value (PPV) and sensitivity are used, remarkable differences can be seen, and will also be discussed in this work.

Sections 5.1, 5.2, and 5.3 show a comparison between different clustering methods for real world applications. While the first two are evaluations performed on sequence clustering of proteins and domains of proteins, the last comparison evaluates the robustness and quality of clustering for PPI networks. For a fair comparison it was chosen to use the same conditions as described in the respective publications. In Section 5.1 this means, that the data sets used in a study by Paccanaro et al. [61] are used as well as the quality measures. Similarly, the analysis in Section 5.3 follows the same strategy as described by Brohée et al. [21]. In addition to these tests, a recently published gold standard is used in the last comparison (Section 5.2), where TC is evaluated against the popular MCL algorithm as well as the recently published Affinity Propagation (AP) approach.

5.1 Single-domain protein sequence clustering

In 2006 Paccanaro et al. [61] performed a comparison of their spectral clustering implementation against other popular clustering tools, namely MCL, GeneRAGE, and hierarchical clustering. The application case here, was to cluster protein domains into related groups. For this task Paccanaro et al. chose to use the SCOP database and, more specifically a subset of the AS-TRAL95 dataset as gold standard. The same data set with the same quality measure, the F-measure, has been used to evaluate the performance of TC. In addition to the original study, the clustering tool AP has also been taken into account.

5.1.1 Data

SCOP is an expert, manually curated database that groups proteins based on their 3D structures. It has an hierarchical structure with four main levels (class, fold, superfamily, family). Proteins in the same class have the same type(s) of secondary structures. Proteins share a common fold if they have the same secondary structures in the same arrangement. Proteins in the same superfamily are believed to be evolutionarily related, whereas proteins in the same family exhibit a clear evolutionary relationship [3]. Here the SCOP superfamily classification is taken as ground truth against which the evaluation of the quality of a clustering generated by a given algorithm is compared. Since the complete SCOP dataset contains many redundant domains that share a very high degree of similarity, most researchers choose to work with the ASTRAL compendium for sequence and structure analysis in order to generate non-redundant data [26]. ASTRAL allows for the selection of SCOP entries that share no more sequence similarity than a given cutoff, removing redundant sequences.

Two subsets of the ASTRAL dataset of SCOP v1.61 are extracted with a cutoff of 95 percent, which means that no two protein domain sequences share more than 95% of sequence identity.

The two subsets are exactly those used in [61]. The first comprises 507 protein domains from 6 different SCOP superfamilies, namely *Globin-like, EF-hand, Cupredoxins, (Trans)glycosidases, Thioredoxin-like,* and *Membrane all-alpha*. This data set is referred to as ASTRAL95_1_161 in the following.

Due to the fact that SCOP is continuously updated, both the original data from [61] (SCOP v1.61) and more recent data from the SCOP version (SCOP v1.71) are evaluated. The novel version is slightly different. For example, the superfamily *Membrane all-alpha* has been removed for the time being, and most of its protein domains are assigned to different superfamilies. Also, several other proteins have been reassigned to one of the five other superfamilies. This provides another dataset of 589 sequences from the remaining 5 superfamilies, which is referred to as ASTRAL95_1_171.

The second subset consists of 511 sequences from 7 superfamilies, namely *Globin-like, Cupredoxins, Viral coat and capsid proteins, Trypsin-like serine proteases, FAD/ NAD(P)-binding domain, MHC antigen-recognition domain,* and *Scorpion toxin-like*, which is referred

5.1. SINGLE-DOMAIN PROTEIN SEQUENCE CLUSTERING

to as ASTRAL95_2_161 and ASTRAL95_2_171 respectively. SCOP can be found at http://scop.mrc-lmb.cam.ac.uk/scop/, while the protein domain sequences are available at http://astral.berkeley.edu/.

5.1.2 Evaluation method

In order to judge the quality of a resulting clustering, Paccanaro et al. chose to use the F-measure (see Section 2.3.1) as quality measure. Note that in the present context, it should not be considered cheating to optimize the similarity function and threshold. The same kind of optimization was applied by Paccanaro et al. in [61]. Although AP was originally not included in this study, it is here to also use recently developed approaches. The same data was used, and the necessary input parameters for AP were optimized to evaluate against the best possible performance of AP. For ASTRAL95_1_161, this was Cov-scoring with $f = 20$ and SoH as secondary scoring function with fixed preference (all self-responsibilities) 600, and damping factor $df = 0.8$. For ASTRAL95_2_161, this was Cov-scoring with $f = 14$ and SoH as secondary scoring function with preference 600, and $df = 0.75$. The results of this comparison have been published in [79] using the predecessor of TransClust, FORCE. Since the underlying model is the same, only the results from that study are presented here.

5.1.3 Results

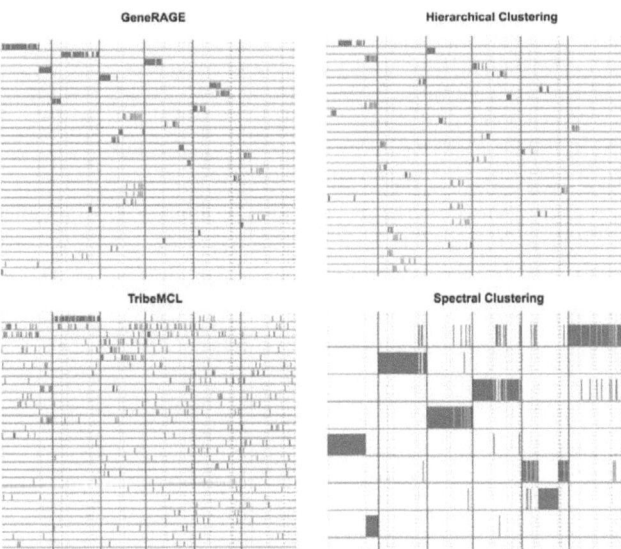

Fig. 5.1: Results for Spectral Clustering, Hierarchical Clustering, MCL, and GeneRage for the ASTRAL95_1_161 dataset. Columns refer to superfamilies and each row refers to a protein domain

5. EVALUATIONS OF THE TRANSITIVITY CLUSTERING MODEL

Table 5.1 summarizes the results: Using FORCE, slightly better agreements than with spectral clustering are obtained. The best similarity function parameters and score threshold for the ASTRAL95_1_161 dataset were Cov-scoring using $f = 20$ and BeH as a secondary scoring function, and $t = -2.2$. For the ASTRAL95_2_161 dataset, this was Cov-scoring with $f = 19$ and SoH as secondary scoring function with $t = -1.6$.

For both datasets, AP performs worse than Spectral clustering. The overall poor performance of AP might be explained by the fact, that AP searches for representatives for a given cluster. This goal is not necessarily appropriate for clustering protein domain sequences.

Dataset	Method	F-measure
ASTRAL95_1_161	FORCE	0.85
ASTRAL95_1_161	Spectral clustering	0.81
ASTRAL95_1_161	Affinity Propagation	0.65
ASTRAL95_1_161	GeneRAGE	0.47
ASTRAL95_1_161	TribeMCL	0.32
ASTRAL95_1_161	Hierarchical clustering	0.26
ASTRAL95_2_161	FORCE	0.89
ASTRAL95_2_161	Spectral clustering	0.82
ASTRAL95_2_161	Affinity Propagation	0.69
ASTRAL95_2_161	GeneRAGE	0.54
ASTRAL95_2_161	TribeMCL	0.52
ASTRAL95_2_161	Hierarchical clustering	0.42

Tab. 5.1: Evaluation of protein clustering tools The F-measure (between 0 and 1) measures the agreement between a clustering resulting from a given algorithm and a reference clustering provided with the dataset. An F-measure of 1 indicates perfect agreement. ASTRAL95_1_161 and ASTRAL95_2_161 refer to the two datasets of SCOP v1.61 used by Paccanaro et al. for spectral clustering [61]. All reported values, except for our algorithm FORCE and for Affinity Propagation, are from the same reference.

Figure 5.2 exemplarily illustrates the clustering results obtained for two similarity functions, and dataset ASTRAL95_1_161. One can see that the classification is very good for the superfamilies *Globin-like*, *EF-hand*, *Cupredoxins*, *(Trans)glycosidases*. *Thioredoxin-like* and *Membrane all-alpha* are split into several clusters. Note, that for *Globin-like* (left column) using similarity function SoH (B), the superfamily is split into two clusters, where the second (the lower one) represents a family.

Additionally FORCE has been applied to the newest ASTRAL95 datasets (ASTRAL95_1_171 and ASTRAL95_2_171). Table 5.2 shows the resulting F-measures for a variety of similarity functions and parameter choices.

5.2 Protein sequence clustering

Finding a good gold standard for remote protein homology detection is difficult, as is the decision of which data to trust and which should be taken into account for an evaluation.

5.2. PROTEIN SEQUENCE CLUSTERING

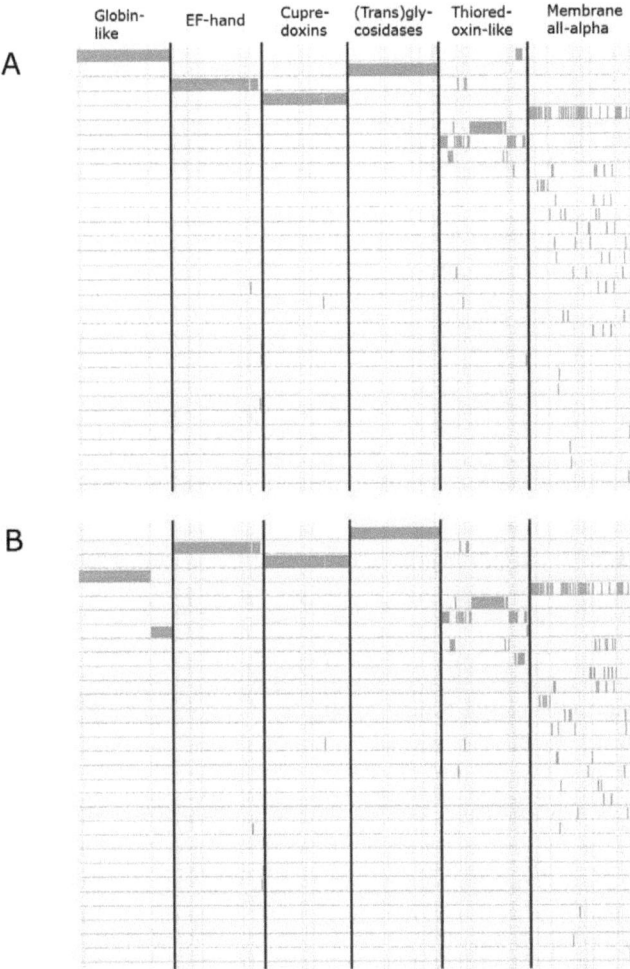

Fig. 5.2: Graphical summary of the obtained clustering results of FORCE for the two similarity functions (A) BeH and (B) SoH, and dataset ASTRAL95_1_161. The MATLAB scripts provided by Paccanaro were used to create images similar to those of Figure 5.1. Each row corresponds to a cluster. Green bars represent a protein assignment to a cluster; each protein is present in only one of the clusters. Boundaries between superfamilies are shown by vertical red lines, and boundaries between families within each superfamily are shown by dotted blue lines.

Dataset	Similarity	Factor f	Threshold	F-measure
ASTRAL95_1_171	SoH	18	-3.0	0.91
ASTRAL95_1_171	BeH	15	-3.4	0.90
ASTRAL95_2_161	SoH	19	-1.6	0.89
ASTRAL95_2_171	SoH	15	-3.2	0.88
ASTRAL95_2_161	BeH	14	-2.4	0.87
ASTRAL95_2_171	BeH	13	-2.6	0.85
ASTRAL95_1_161	BeH	20	-2.2	0.85
ASTRAL95_1_161	SoH	20	-1.8	0.83

Tab. 5.2: Evaluation of the WGCEP model The best F-measures for each dataset and each similarity function. ASTRAL95_1_161 and ASTRAL95_2_161 are as in Table 5.1. ASTRAL95_1_171 and ASTRAL95_2_171 refer to the updated ASTRAL95 data of SCOP v1.71. BeH or SoH denote the similarity function, while the coverage factor f represents the influence of the coverage to the similarity.

Most of the comparative studies presented here are based on previous experiments. In addition to this, a dataset created by Brown et al. [22] is used as an independent evaluation. This choice of data is based on the fact that it was specifically created and hand curated as a gold standard dataset. Three different experiments are performed using this data; a comparison to different clustering methods as in the previous evaluation, an example of the density parameter estimation routine of TransClust, and an evaluation of the impact of integrating additional information on the quality of a clustering.

5.2.1 Data

For the experiments in this section, a dataset published by Brown et al. [22] is used, consisting of 866 enzymes that were manually assigned to protein families and superfamilies. Since this dataset is hand curated it is well-suited as a gold standard.

First similarities were calculated between the protein sequences using all-vs.-all BLAST results. In contrast to the domain clustering, the score similarity function (see Section 2.2.2) defines the pairwise similarity throughout this study. This similarity function requires the restriction of the High Scoring Pairs (HSPs) of a BLAST file to those with low E-values since, otherwise, the normalization of the score would increase even 'bad' HSPs to be equally valued as those with low expectation value. The best results could be obtained using a cutoff of $1 \cdot 10^{-5}$, wherein only HSPs with E-value lower than this threshold are taken into account.

5.2.2 Comparison to different clustering methods

In order to ensure a fair comparison between the clustering approaches TC, MCL, and AP all parameters were again optimized and only the best results are compared. The range of the corresponding density parameters can be found in Table 5.3.

The F-measure (refer to Section 2.3.1) is used as the quality measure, since it gives a good idea of the accuracy of the resulting clusterings.

5.2. PROTEIN SEQUENCE CLUSTERING

Method	Density parameter	Min value	Max value	Step size
TC	threshold	0	1	0.05
AP	preferences	-10	2	0.1
MCL	inflation factor	1.1	5	0.01

Tab. 5.3: Summary of the used density parameters for the protein clustering evaluation.

Method	best density parameter	best F-measure
TC	0.4	0.93
AP	-2.9	0.67
MCL	2.15	0.89

Tab. 5.4: List of optimal density parameters with corresponding F-measure for protein clustering.

Although the differences are not as high as in the previous experiment, one can see from Table 5.4 that TC outperforms the other two methods. As in the last section, the relatively bad results of AP may be explained due to its goal. Searching for perfect representatives for the clusters does not reflect the group attributes of a protein cluster. MCL performs much better than in the domain superfamily identification, which indicates that it is much more suitable for identifying smaller groups than large ones.

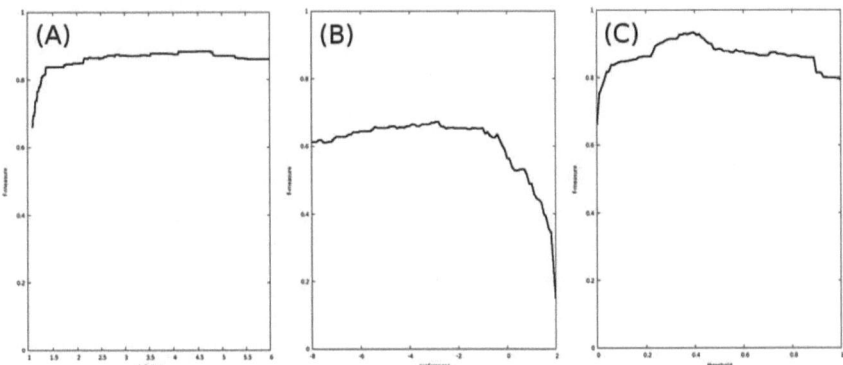

Fig. 5.3: Illustration of the impact of the choice of the density parameter on the quality of the resulting clustering. (A) for MCL with varying inflation factors, (B) for AP with varying initial self-responsibilities (preferences), (C) for TC with varying threshold

5.2.3 Example threshold determination

Figure 5.3 shows the impact of the chosen density parameter on the resulting clustering. As expected the F-measure changes for different choices of density parameters for all three approaches. This illustrates well how important the density parameter for each clustering al-

5. EVALUATIONS OF THE TRANSITIVITY CLUSTERING MODEL

gorithm is, and hence how much a method for the detection of such a parameter is needed. TransClust integrates such methods that aid in finding a good threshold. As illustrated in Figure 5.5 one can use the Cytoscape plugins for guessing a good range for the similarity threshold. An example for such an analysis is given below.

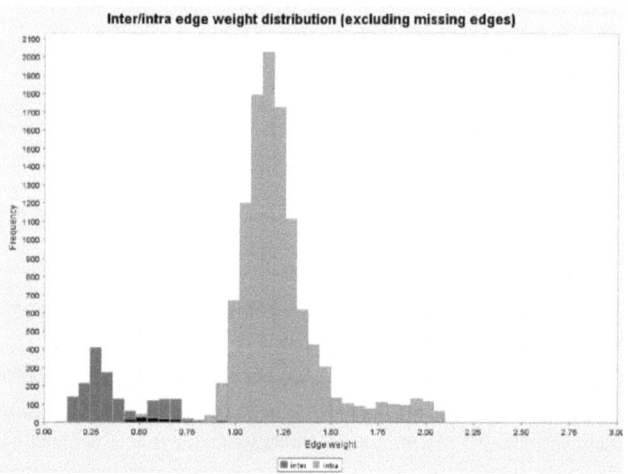

Fig. 5.4: Screenshot of the inter vs. intra edge weight distribution of the amidohydrolase superfamily using the ClusterExplorer Cytoscape plugin. The similarity function is the normalized bit-score

(i) The dataset is divided into two subsets, a training set consisting of all 232 proteins of the amidohydrolase superfamily, and a test set of the remaining 634 proteins. Two according similarity files and two gold standard files, containing only the elements of the respective two sets, are created from the original files of the whole dataset. The original similarity file was built using the score similarity function and a blastcutoff of $1 \cdot 10^{-5}$.

(ii) Using the Cytoscape plugin "ClusterExplorer" the inter versus intra edge weight distribution of the training set has been plotted (see Figure 5.4). Using this as reference a good threshold can be found between 0.4 and 0.8, and no inter edges can be found with weight higher than 1.

(iii) The clustering for the training set with thresholds of the range, described above and an upper bound of 1 are calculated and compared to the gold standard file of this subset. The highest F-measure of 0.977 can be found at threshold 0.48.

(iv) The best threshold is applied to the test set and in order to evaluate the quality of this prediction the F-measure is calculated. At threshold 0.48 an F-measure of 0.892 can be achieved, while the best threshold for this set would be 0.42 with an F-measure of 0.933.

5.3. CLUSTERING PROTEIN-PROTEIN INTERACTION NETWORKS   79

One can see from this example that although the best value has not been found, this procedure helps in guessing a good threshold. The visualization of the edge weight distribution via the corresponding Cytoscape plugin gives an initial intuition about the range, where to search the threshold. Using this initial information, the threshold can be determined more precisely by clustering the respective test data. More experiments would be necessary to sufficiently evaluate this method, but to do so, alternative and bigger gold standard sets are necessary.

5.2.4 Integration of additional knowledge

TC allows for the integration of additional knowledge. Such information can consist of a known assignment for a subset or an upper bound, as used in the previous example. An experiment should demonstrate how the integration of additional knowledge can drastically improve the clustering results. From the gold standard assignment, it is known which proteins should be in the same cluster and which should be in different groups. If only a fraction of these pairwise assignments is known, i.e. if two proteins belong to the same or different clusters, this already helps in avoiding false assignments. In the following study, a certain percentage out of all pairwise assignments is considered as known. Randomly these pairwise assignments are picked and the corresponding similarity between the two elements is set to 100,000 if the nodes are in the same gold standard cluster, and to -100,000, if not. Merging all nodes above the upper bound 10,000 guarantees that two elements from the same cluster cannot be separated. The results can be seen in Figure 5.6 for varying percentage of knowledge. The experiment is repeated 10 times for each percentage and the average of the best F-measures is displayed. Often an assignment for subsets is known. This is simulated by only integrating information about elements within a gold standard cluster. Out of all edges between two elements of one cluster, a certain percentage is randomly chosen and set to 100,000.

One can see from Figure 5.6 that even if only a few interactions are known, this already improves the clustering results drastically. Further tests may be necessary, but this experiment clearly shows how important it is to have the ability to integrate additional information into a clustering process.

5.3 *Clustering protein-protein interaction networks*

In 2006, Brohée et al. [21] presented an evaluation of four graph based clustering algorithms for the task of reconstructing PPIs. The algorithms compared were MCL, RNSC [49], MCODE [7], and SPC [19]. In order not to replicate existing results, this work concentrates only on the best algorithms of this experiment, namely MCL and RNSC. A robustness analysis compares the performance of the clustering approaches on altered graphs, where edges are removed and added at random. Large scale experiments provide real-world data, on which the different approaches are applied again. Furthermore, this study depicts the impact of the choice of the quality measure on the resulting clustering. The last experiment in this section is similar to the robustness analysis, but evaluates the performance of two presented overlapping methods

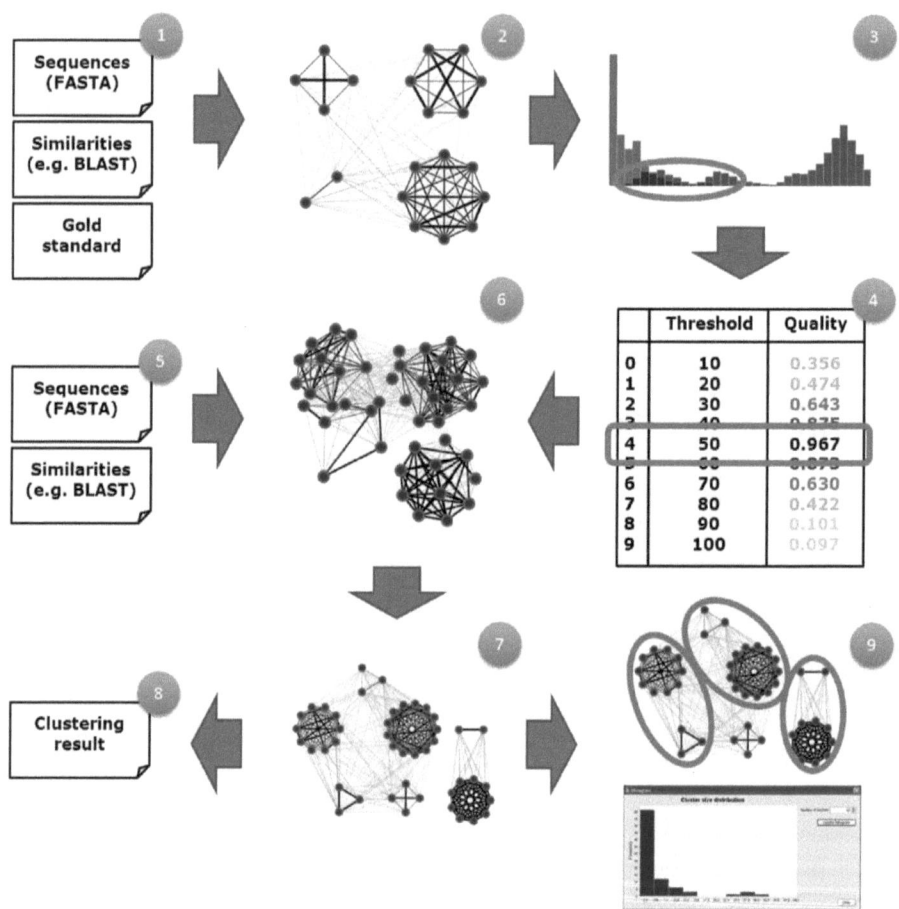

Fig. 5.5: The following steps can be performed assisted by TransClust and Cytoscape. (1) Import of the amino acid sequences and all-vs.-all BLAST results, as well as the known cluster assignment for the gold standard subset. (2) Computation and visualization of the corresponding similarity network. In the figure, edge color and thickness correlate with the assigned similarity values. (3) Plotting of the intra- vs. inter-cluster similarity distribution for the gold standard clusters to estimate a promising region for the density parameter, i.e. the similarity threshold (see Supporting Information). (4) Iterative clustering with varying thresholds and comparison of the clustering results with the gold standard to identify the "best threshold" that reconstructs the gold standard (quality). (5) Import of the full data set. (6) Visualization of the corresponding full similarity network and (7) clustering of this network assuming that the threshold is conserved between the gold standard subset and the full data set. (8) Export the results or (9) perform further analyses by using ClusterExplorer

5.3. CLUSTERING PROTEIN-PROTEIN INTERACTION NETWORKS

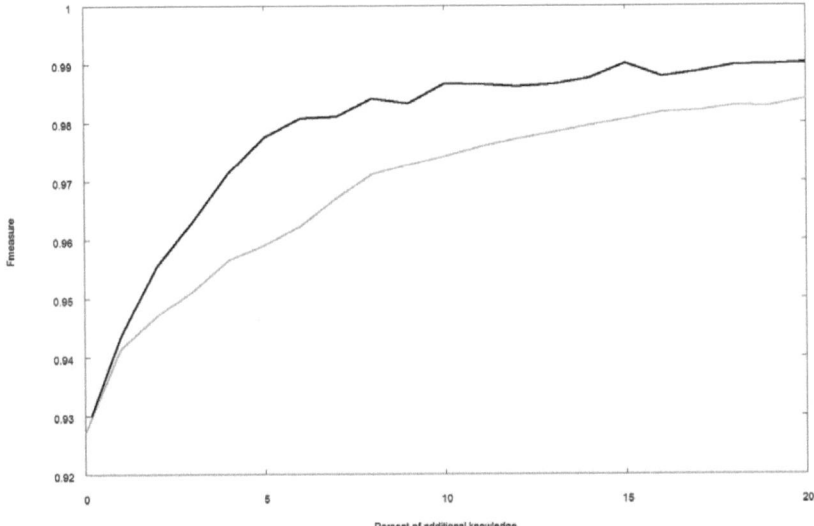

Fig. 5.6: Illustration of the impact of the amount of additional knowledge added for all pairwise interactions (red curve) and all interactions within gold standard groups (green curve) based on TC.

5.3.1 Data

This evaluation uses the same data as in [21] to ensure a fair comparison. The first dataset consists of 1095 proteins of the yeast *Saccharomyces cerevisiae* obtained from the MIPS database [58]. The complex annotations of these proteins are used throughout the whole evaluation as the gold standard. For a robustness analysis, it is started with an initial graph $G = (V, E)$ where V represents the proteins, and an interaction is assumed, and thus an edge is drawn between every two proteins of a complex. Note that a protein can belong to multiple complexes, which makes the graph intransitive (refer to Figure D.6 A for a visualization of this graph). In the following, this graph is modified by randomly adding and deleting a certain amount of edges. This has been done in [21] and the modified graphs are available at http://rsat.bigre.ulb.ac.be/rsat/data/published_data/brohee_2006_clustering_evaluation/. In what follows, let $A_{i,j}$ denote the graph where $i\%$ of edges are added and $j\%$ of edges are deleted (refer to Figure D.6 B for the graph $A_{100,40}$ with 100% added and 40% deleted edges).

Furthermore, the performance of the different approaches was tested with high throughput experiments obtained from the GRID database [20]. Two different methods have been used to detect PPIs. Gavin *et al.* [37, 38], Ho *et al.* [44], and Krogan *et al.* [52] used mass spectrometry while Uetz *et al.* [75] and Ito *et al.* [45] predicted interactions using the two-hybrid technique. A summary of the data can be found in Table 5.5, where the results of the different clustering algorithm on these data set are also displayed.

As a negative control and reference for the quality measure, the protein names in the gold standard file were shuffled.

5.3.2 Robustness analysis

The evaluation presented here is a robustness analysis equal to those of MCL and RNSC as described in [21]. The clusters calculated with TransClust were obtained by using thresholds between 0 and 1 with steps of 0.05. Results for all graphs $A_{i,j}$ of the original study were compared against the gold standard. For this evaluation, the F-measure serves as quality measure in addition to the originally used accuracy and separation. The results from the original study and the performance of TC measured with separation and accuracy can be found in the appendix. Each altered graph was tested with a series of density parameters for all approaches. In order to test the robustness, it is not the best value that is presented, but rather the values from that density parameter that lead to the best results in the most cases. Figure D.7 illustrates the results from the original study as a reference. The analysis has been repeated for TC. Figure 5.7 illustrates the results for all altered graphs using the F-measure as quality measure.

TC and RNSC produce similar results for the tested graphs. Both methods are robust against edge additions and quite robust against edge deletions. MCL generally has lower F-measures and seems also to be less robust against edge additions. All tested methods show a clear difference between the altered graphs and those that have additionally been shuffled as a control group.

5.3. CLUSTERING PROTEIN-PROTEIN INTERACTION NETWORKS

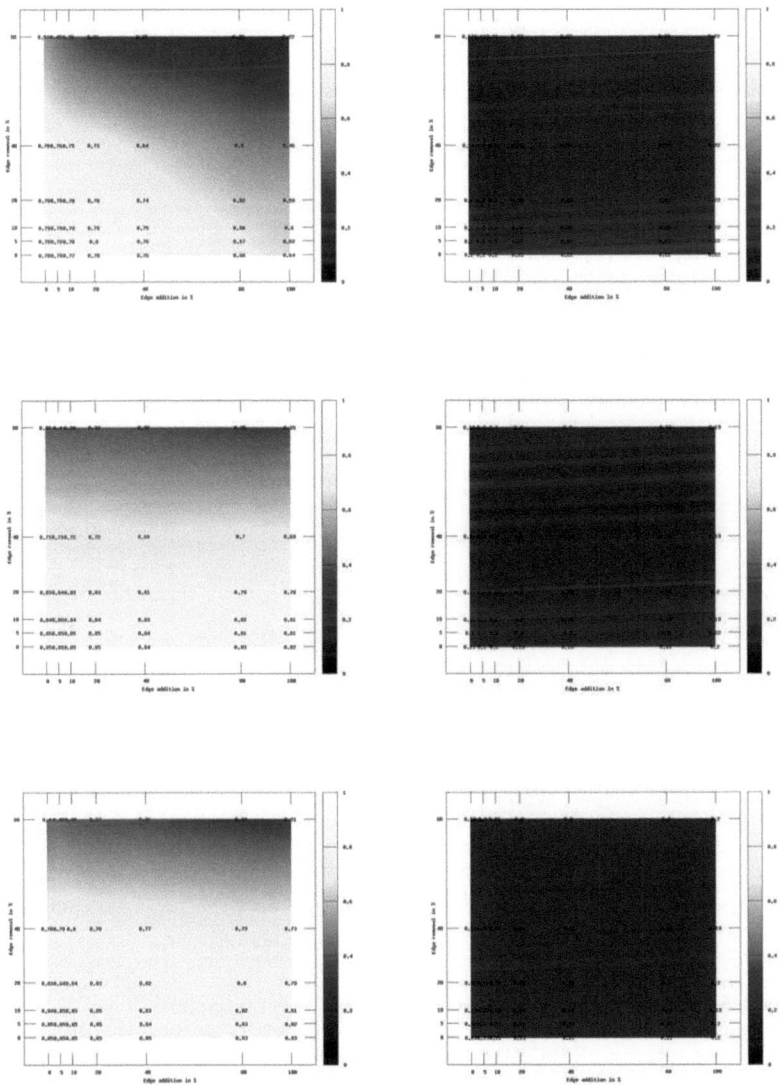

Fig. 5.7: Illustration of the robustness analysis. On the left side the results for the altered graphs is displayed and on the right side one can see the results for the negative control group. The F-measure serves as quality measure. Results are shown for MSL (top), RNSC (middle), and TransClust (bottom).

5.3.3 Evaluation on experimental data

In order to compare TC, the same quality measures as in [21] are used, namely the PPV, the sensitivity, the geometric and arithmetic accuracy, as well as the separation (refer to Section 2.3.1 for a detailed description of this measurements). In addition to this, the best methods MCL and RNSC are again applied on all used datasets and the F-measure is calculated. These additional experiments give interesting insights into how important the choice of the quality function is and how it influences the results.

Dataset	Measure	MCL		MCODE		RNSC		SPC		TC	
		orig	rand	orig	rand	orig	rand	orig	rand	orig	rand
Uetz et al.											
926 nodes	Sn	57.3	38.6	84.3	74.5	49.4	36.5	65.5	43.3	52.05	36.55
865 edges	PPV	53.8	45.9	25.5	21.6	59.6	54.4	38.0	38.9	57.89	54.97
1.175 mean	ACC	55.6	42.3	54.9	48.0	54.5	45.5	51.8	41.1	54.97	45.76
degree	Sep-co	23.0	20.6	48.9	62.5	15.5	14.8	19.1	21.2	18.33	18.26
	Sep-cl	30.1	26.9	2.2	2.8	34.3	32.7	20.3	22.6	33.17	32.87
	Sep	26.3	23.5	10.4	13.3	23.1	22.0	19.7	21.9	24.66	24.50
Ito et al.		orig	rand	orig	rand	orig	rand	orig	rand	orig	rand
2937 nodes	Sn	34.9	26.0	66.9	68.0	31.4	24.0	73.2	64.6	32.31	24.35
4038 edges	PPV	42.7	38.5	8.2	5.8	63.6	61.8	24.3	23.8	62.39	60.07
2.682 mean	ACC	38.8	32.2	37.5	36.9	47.5	42.9	48.8	44.2	47.35	42.21
degree	Sep-co	12.7	11.8	41.6	33.0	7.1	7.0	11.3	11.0	8.30	54.13
	Sep-cl	36.2	33.9	1.7	1.3	56.7	55.9	20.1	20.4	55.61	54.13
	Sep	21.4	20	8.4	6.7	20.1	19.8	15.4	15.0	21.13	21.2
Ho et al.		orig	rand	orig	rand	orig	rand	orig	rand	orig	rand
1352 nodes	Sn	50.6	28.2	81.2	76.5	37.0	27.4	90.1	92.1	41.19	26.92
3210 edges	PPV	47.1	35.6	12.9	8.5	61.5	57.1	10.4	8.2	58.01	50.80
4.7 mean	ACC	48.9	31.9	47.1	42.5	49.3	42.2	50.2	50.2	49.60	38.86
degree	Sep-co	22.6	19	44.7	37.2	11	10.5	19.3	13.8	13.33	12.95
	Sep-cl	32.3	27.1	2.6	2.2	48	45.6	5.5	4.0	45.15	40.67
	Sep	27.0	22.7	10.9	9.0	20	21.0	10.3	7.4	24.53	22.95
Gavin et al.		orig	rand	orig	rand	orig	rand	orig	rand	orig	rand
2002	Sn	74.1	24.2	67.0	51.1	52.1	20.8	91.8	81.4	57.98	20.81
1352 nodes	PPV	57.0	23.9	20.4	9.4	62.0	46.0	18.1	10.7	60.24	42.93
3210 edges	ACC	34	24.0	43.7	30.3	57.1	33.4	54.9	46.0	59.11	31.87
4.7 mean	Sep-co	39.4	17.6	44.5	16.1	14.5	11.3	34.4	15.7	15.44	14.77
degree	Sep-cl	38.0	17.0	5.5	2.0	46.9	36.5	13.6	6.2	42.81	35.18
	Sep	38.7	17.3	15.6	5.6	26.1	20.3	21.6	9.8	25.71	22.79
Gavin et al.		orig	rand	orig	rand	orig	rand	orig	rand	orig	rand
2006	Sn	75.7	23.7	58.3	43.2	60.8	20.9	79.8	48.4	64.59	19.84
1430 nodes	PPV	54.3	21.0	20.6	8.0	63.3	37.3	37.0	16.5	62.71	45.23
6531 edges	ACC	65.0	22.4	39.5	25.6	62.1	29.1	58.4	32.4	63.65	32.54
9.1 mean	Sep-co	38.1	15.5	44.7	15.3	20.1	12.9	34.9	14.9	20.89	14.73
degree	Sep-cl	32.7	13.3	7.9	2.7	44.5	28.6	21.6	9.2	42.34	35.29
	Sep	35.3	14.4	18.8	6.4	29.9	19.2	27.4	11.7	29.74	22.8
Krogan et al.		orig	rand	orig	rand	orig	rand	orig	rand	orig	rand
2675 nodes	Sn	62.8	19.8	56.3	30.9	53.1	19.1	82.6	64.0	56.47	18.85
7088 edges	PPV	56.2	33.5	21.9	9.7	63.3	51.1	25.4	17.2	62.29	50.00
5.296 mean	ACC	59.5	26.7	39.1	20.3	58.2	35.1	54.0	40.6	59.38	34.43
degree	Sep-co	20.0	12.1	33.2	13.6	10.3	8.7	20.3	11.9	11.99	10.82
	Sep-cl	49.5	29.9	8.8	3.6	59.6	50.3	24.0	14.1	58.00	48.77
	Sep	31.5	19.0	17.0	7.0	24.7	21.6	20.9	12.9	26.37	22.97

Tab. 5.5: Comparison between the different clustering approaches from Brohée et al. [21] including results for TransClust. Abbreviations: MCL: Markov Clustering, MCODE: Molecular Complex Detection, RNSC: Restricted Neighborhood Search Clustering, SPC: Super Paramagnetic Clustering, TC: Transitivity Clustering, orig: original dataset, rand: control group with random changes of the gold standard assignments, Sn: Sensitivity, PPV: Positive Predictive Value, ACC: geometric Accuracy, Sep-cl: cluster-wise Separation, Sep-co: complex-wise (reference-cluster-wise) Separation, Sep: Separation. Parts taken from [21].

In the original experiment the quality measures were calculated by comparing the clustering results against the complete gold standard, as used in the robustness analysis. This explains most of the "bad" results for data sets like "Uetz et al.", for instance, where only a fraction

5.3. CLUSTERING PROTEIN-PROTEIN INTERACTION NETWORKS 85

of the clusters can be found (refer to Table 5.5 for the results of the original study and TC). Consequently the following comparison restricts the gold standard to those elements that intersect with the respective data set and, moreover, also deletes any element from a cluster that cannot be found in the gold standard. This restricts the experiment to those elements where knowledge is actually available. Also a different quality measure, the F-measure, is used to get an additional point of view on the results. The separation, which serves as the main quality measure in the original study, strongly depends on the number of clusters. An approach that produces a lot of singletons and may be more "careful" with an assignment cannot reach a high score. The F-measure compares a gold standard group with the best cluster in the produced clustering and hence ignores this drawback. The difficulties in interpreting the results can also be seen by comparing the gold standard data set against itself. Using the quality measures presented by Brohée *et al.* this leads to values of 0.62 for the separation and 0.57 for the accuracy, which are smaller than the results achieved by clustering methods. The F-measure is 1 for this comparison, and always smaller when using a clustering result.

Table 5.6 summarizes the results for MCL, RNSC, and TC as applied to the large scale data sets, where the quality measures are applied on the restricted gold standard and clustering result set. The density parameters were optimized for a best possible F-measure for each of the methods. MCLs inflation value varies from 1.1 to 5, RNSCs density parameter (maximal number of clusters) varies from 1 to 3000, and TCs threshold varies from 0.1 to 1.

One can clearly see the differences between the results from the original study and the results from the new study. Restricting the evaluation to those elements that occur in both sets, the real-world data and the gold standard data set, leads to higher values in accuracy and separation. MCL still performs well, but the difference in quality to the other methods is much smaller, and sometimes RNSC and TC even outperform it. The difference between the methods in the second analysis is in general much smaller than in the original study. If confidence information about the interactions would be available, it is expected that methods like MCL and TC perform even better, since they are designed for incorporating weights, in contrast to RNSC, which is only defined for unweighted problems. In such cases, one could again take advantage of the intuitive density parameter of TransClust.

Dataset	Measure	MCL		RNSC		TC	
		orig	rand	orig	rand	orig	rand
Gavin 2002	Density	1.8	1.8	271	271	0.1	0.1
	F-measure	72.71	27.16	69.07	26.99	70.42	26.93
	PPV	57.01	33.85	57.13	29.69	56.65	29.93
	Sn	74.11	21.26	64.85	21.26	70.43	21.62
	ACC	65.00	26.82	60.86	25.12	63.16	25.43
	Sep-co	62.38	21.47	48.96	21.94	50.87	21.65
	Sep-cl	49.46	34.93	50.40	31.54	51.17	30.88
	Sep	55.54	27.38	49.68	26.30	51.02	25.86
Gavin 2006	Density	2	2	301	301	0.3	0.3
	F-measure	68.67	27.93	68.75	25.32	68.60	27.25
	PPV	56.59	49.50	60.02	32.00	62.79	45.29
	Sn	72.76	20.38	64.45	20.38	64.67	19.82
	ACC	64.17	31.76	62.20	25.54	63.73	29.96
	Sep-co	52.37	17.91	44.76	19.45	41.12	19.37
	Sep-cl	46.25	45.98	51.82	31.28	55.06	45.43
	Sep	49.21	28.70	48.16	24.66	47.58	29.67
Ho	Density	2.4	2.4	231	231	0.1	0.1
	F-measure	55.52	34.95	55.67	32.33	55.32	33.34
	PPV	53.53	45.03	46.15	30.93	49.52	37.34
	Sn	48.40	26.76	51.12	28.04	49.84	27.24
	ACC	50.90	34.72	48.57	29.45	49.68	31.89
	Sep-co	39.11	28.08	39.71	28.28	40.40	28.27
	Sep-cl	51.76	47.25	40.42	31.50	46.93	38.94
	Sep	44.99	36.43	40.06	29.85	43.54	33.18
Ito	Density	2.6	2.6	971	971	0.15	0.15
	F-measure	41.70	32.01	43.62	32.32	43.40	33.04
	PPV	52.90	51.79	54.99	53.27	59.31	55.98
	Sn	32.92	25.15	34.16	24.29	33.79	24.29
	ACC	41.73	36.09	43.34	35.97	44.76	36.88
	Sep-co	29.20	24.78	29.43	24.42	27.77	25.14
	Sep-cl	54.01	52.78	55.81	54.29	58.84	58.74
	Sep	39.71	36.17	40.53	36.41	40.42	38.43
Krogan	Density	1.8	1.8	421	421	0.1	0.1
	F-measure	63.99	27.32	64.99	25.68	64.97	26.20
	PPV	56.19	45.10	57.86	33.09	56.84	37.06
	Sn	62.85	19.41	60.44	19.50	64.42	19.22
	ACC	59.43	29.59	59.14	25.40	60.51	26.69
	Sep-co	43.07	19.40	42.52	19.64	44.80	20.13
	Sep-cl	54.22	46.32	54.78	33.81	56.39	39.35
	Sep	48.33	29.97	48.26	25.77	50.26	28.14
Uetz	Density	2.3	2.3	261	261	0.15	0.15
	F-measure	62.34	45.29	63.11	44.27	62.54	45.11
	PPV	55.85	50.00	53.80	45.61	56.14	47.66
	Sn	56.14	37.13	58.77	37.43	54.97	37.43
	ACC	55.99	43.09	56.23	41.32	55.55	42.24
	Sep-co	49.28	38.37	49.69	38.00	49.38	38.19
	Sep-cl	55.59	52.19	52.48	46.81	56.49	49.19
	Sep	52.34	44.75	51.07	42.17	52.82	43.35

Tab. 5.6: Comparison between MCL, RNSC, and TC on large scale data. The density parameters were optimized for a best F-measure. Interesting values are highlighted; F-measure (red), accuracy (blue) and separation (green). Abbreviations: MCL: Markov Clustering, RNSC: Restricted Neighborhood Search Clustering, TC: Transitivity Clustering, orig: original dataset, rand: control group with random changes of the gold standard assignments, Sn: Sensitivity, PPV: Positive Predictive Value, ACC: geometric Accuracy, Sep-cl: cluster-wise Separation, Sep-co: complex-wise (reference-cluster-wise) Separation, Sep: Separation.

5.3. CLUSTERING PROTEIN-PROTEIN INTERACTION NETWORKS

5.3.4 Finding overlaps with Transitivity Clustering

The gold standard set has an overlapping structure, which makes it perfect to evaluate the corresponding modifications to TC as presented in Section 3.3.3. The first method used in this evaluation uses average similarities between objects and other clusters to assign them to additional groups. For this method a second threshold between 0 and 1 is necessary. The following Table 5.7 includes results for a threshold of 0.8. The second method used here assigns objects to additional clusters, providing that reduces the internal costs for that cluster. No additional value is necessary for this method.

For this study, the F-measure serves as the quality function. The datasets used for this evaluation are the altered graphs, which were also used in the robustness analysis. The average of the density parameters that lead to the best results for each of the altered graphs is used to create the subsequent results. This restriction to the average threshold also helps to evaluate the robustness of the overlapping clustering methods.

	0	05	10	20
0	0.85/0.87/0.87	0.85/0.86/0.86	0.85/0.85/0.86	0.85/0.85/0.86
5	0.85/0.86/0.87	0.85/0.86/0.86	0.85/0.85/0.86	0.85/0.85/0.85
10	0.84/0.85/0.86	0.85/0.85/0.86	0.85/0.85/0.86	0.85/0.85/0.85
20	0.83/0.83/0.85	0.84/0.84/0.85	0.84/0.84/0.84	0.83/0.83/0.84
40	0.78/0.79/0.81	0.79/0.8/0.81	0.8/0.8/0.81	0.78/0.79/0.79
80	0.4/0.42/0.44	0.42/0.43/0.45	0.39/0.43/0.41	0.37/0.39/0.4

	40	80	100
0	0.85/0.85/0.85	0.83/0.84 /0.83	0.83/0.82/0.82
5	0.84/0.84/0.84	0.83/0.83/0.82	0.82/0.82/0.81
10	0.83/0.83/0.84	0.82/0.82/0.82	0.81/0.81/0.81
20	0.82/0.81/0.83	0.8/0.8/0.8	0.79/0.79/0.79
40	0.77/0.78/0.77	0.73/0.74/0.72	0.73/0.73/0.7
80	0.36/0.36/0.37	0.32/0.31/0.32	0.31/0.3/0.31

Tab. 5.7: Comparison of the overlapping methods of TC. F-measures for the original partitional clustering (black), the overlapping methods using either fuzzy associations (red) or direct secondary assignments (blue). The columns represent the amount of edges that were added while the rows are the percent of removed edges.

Table 5.7 compares the results of the different methods. Both overlapping methods perform better than the original partitional TC approach. This is only a limited evaluation but it illustrates already that these methods can be applied to produce overlapping clusters.

6. INTEGRATED APPLICATIONS

TransClust, the implementation of TC, allows for an easy integration into existing software or calculation pipelines. The aforementioned integration into the Cytoscape framework is only one example. Additionally, TC has been successfully used as part of the MoRAine program [16], a method for increasing the information content of a Position Specific Scoring Matrix (PSSM). Another application has been the integration of TC into a pipeline to transfer transcriptional gene regulatory networks from a model organism to closely related species. Since 2007, TC has been part of CoryneRegNet [8, 14], a reference database for corynebacterial gene regulatory networks, where it is utilized for the identification of clusters of homologous proteins based on their sequence similarity.

6.1 MoRAine

In this section it will be shown that TC is applicable for optimizing position specific scoring matrices of transcription factor binding motifs. The goal here is to increase the subsequent performance of prediction methods for finding new binding motifs. The related framework MoRAine 1.0 was published in 2007 [16]. MoRAine 1.0 utilized several clustering methods for this task. An improved version of this program, MoRAine 2.0, now integrates TC as clustering method. In the following subsections, the software MoRAine and its underlying problem will be described.

6.1.1 Transcription factor binding site annotation - A difficult and error-prone task

The so-called transcription factors are important components of the cell's regulatory machinery. They are DNA-binding proteins that are able to detect intra- and extracellular signals. By binding to so-called Transcription Factor Binding Motifs (TFBMs) they control the expression of their target genes, thereby decisively influencing genetic programs like growth, reproduction, and defense [5, 6, 14, 60, 66]. Given a set of known TFBMs for a certain regulator, one can compute mathematical models to perform *in silico* predictions of further TFBMs in order to predict regulatory networks. This task is generally complicated by the relatively low level of TFBM conservation. The most widely used model for TFBMs are so-called Position Frequency Matrices (PFMs) [72]. PFMs can be converted to Position Specific Scoring Matrices (PSSMs) by calculating log-odds scores. These matrices are used, in turn, to predict TFBMs in the upstream sequences of putative target genes for a certain Transcription Factor (TF). Various software tools are available: PoSSuMsearch [17], Virtual Footprint [59], MATCH [48], and

P-MATCH [28], just to name a few.

Nowadays, TFBM wet lab determination is done by electrophoretic mobility shift assays (EMSA) [43], DNAse footprinting [36], ChIP-chip [73], ChIP-seq [47], or mutations of putative TFBMs and subsequent expression studies. All of these methods lack a precise binding sequence identification that is accurate to one base pair [12]. Another problem occurs: Since TFs bind the double-stranded DNA, the question of which strand of the TF-binding sequence is annotated becomes a matter of interpretation. Clearly, both issues directly affect and complicate TFBM modeling as position frequency matrices and hence, the subsequent PSSM-based binding site prediction. This problem occurs when a TFBM from either strand, based on approximate knowledge of its position, is entered in a reference database and subsequently used blindly for PSSM-based predictions. This does happen in practice for regulatory databases that integrate information from other sources [16], for instance, in CoryneRegNet [8, 10, 12].

For mis-annotated TFBMs, one may observe a poor information content of the subsequently computed PFM, which consequently leads to a decreased binding motif prediction for the PSSM that was constructed from that PFM. This problem can be solved by re-annotating the TFBMs by possibly switching their strands and/or shifting them a few positions, in order to maximize the information content of the resulting PFM.

6.1.2 Methods

The following definitions are needed to explain how MoRAine works and to compare the re-adjustment performances of MoRAine 1.0 and 2.0.

Let $\Sigma := \{A,T,C,G\}$ be the DNA alphabet. In accordance with [16], a position frequency matrix $F = (f_{\sigma j})$ for a set of n TFBMs of length m over the alphabet Σ is defined as a $|\Sigma| \times m$ matrix, where $f_{\sigma j}$ is the relative frequency of letter σ at position j.

Crooks et al. introduced in [30] the information content as quality measure for PFMs. The information content I_j for column j of F is defined as

$$I_j := \log_2 |\Sigma| + \sum_{\sigma \in \Sigma} f_{\sigma j} \cdot \log_2 f_{\sigma j} \quad \text{[bits]}.$$

If all symbols at position j agree, I_j reaches its maximum with maximal value 2 bits for a 4-letter alphabet Σ. The mean information content $I(F)$ for a given PFM F is defined as the average I_j over all positions j:

$$I(F) := \frac{1}{m} \sum_{j=1}^{m} I_j.$$

In what follows, we use the mean information content $I(F)$ as a quality measure for a given PFM F and denote it shortly with I if F is fixed. The information content is used to compare the quality of two different PFMs F_1 and F_2 by comparing $I(F_1)$ with $I(F_2)$. If F_2 is the PFM of the MoRAine-adjusted TFBMs, while F_1 is the PFM computed from the input TFBMs, with $I(F_1) \leq I(F_2)$, one can calculate the percentage improvement performance P with $P = 100 \cdot \frac{I(F_2)}{I(F_1)}$.

6.1. MORAINE

MoRAine now works as follows: The input is a set of n annotated length-m TFBM sequences that extend l bp to the left and r bp to the right. Hence, the length of the given input sequences is $m^+ := m + l + r$. First, MoRAine computes the set M of every possible motif of length $m = m^+ - l - r$ derived by the operations *shift* and *switch* applied to each of the n input sequences. The operation *shift* provides every substring of length m for a given motif of length m^+, and the operation *switch* its reverse complement sequence. We obtain a set S_i of $M := |S_i| = 2 \cdot (l + r + 1)$ potential TFBM sequences of length m for each input sequence i, with $i = 1, \ldots, n$.

So far MoRAine 1.0 and 2.0 work in a similar way. For both, the goal is to find a set C of TFBMs that contains exactly one TFBM from each S_i and maximizes the mean information content of the corresponding PFM F_C. MoRAine 1.0 offers two heuristic clustering algorithms, (cg) and (km), both working on either of two similarity functions, $(simC)$ and $(simS)$ (a description of these methods can be found in the Appendix). Table 6.1 summarizes the running times and TFBM annotation improvement performance of MoRAine 1.0 for all four combinations. One can see a trade-off between accuracy and running time: $(cg/simS)$ provides best results but $(cg/simC)$ is much faster.

Using TC as the clustering method in MoRAine 2.0 closes this gap and provides a powerful tool that now provides better results than MoRAine 1.0 with $(cg/simS)$ at running times equal to $(cg/simC)$. The goal can be cast as follows: Partition the set of input TFBMs into $M = 2 \cdot (l + r + 1)$ clusters, where each cluster contains exactly n motifs, one of each S_i ($i = 1, \ldots, n$) and thus is a putative solution. In the following, it is described how TC was integrated with MoRAine 2.0 to find such a set C.

TC is flexible and offers the capability to integrate additional knowledge, making it perfect for the needs here. As similarity function, instead of the functions from MoRAine 1.0, $(simC)$ and $(simS)$, the difference between the motif length $l = |p| = |q|$ and the hamming distance $h(p,q)$ for two TFBMs p, q: $s(p,q) = l - h(p,q)$ is used. To ensure that each cluster of TFBMs contains only one motif from each set S_i, the similarity function s is set to $-\infty$ if $p \in S_i$ and $q \in S_i$ for any S_i, i.e. if both potential solutions (the TFBMs p and q) originate from the same input TFBM. The threshold t is set to zero, which guarantees that each cluster contains exactly one TFBM from each set S_i. TC's integration with MoRAine is mainly responsible for the increased performance of MoRAine 2.0, as will be demonstrated in the following section.

6.1.3 Results and discussion

Implementation

MoRAine 2.0 is an open source JAVA 6 program. It can be accessed and downloaded at http://moraine.cebitec.uni-bielefeld.de. As shown for MoRAine 1.0 in [13], release 2.0 of MoRAine may be included into a database back-end as a quality assurance tool, or to provide a bioinformatics workflow with adjusted position weight matrices for TFBM predictions.

MoRAine 2.0 can still be used as web application. The user may copy and paste binding sequences in FASTA format at the MoRAine web site to calculate the adjusted motifs as well as the corresponding sequence logos by using the Berkeley web logo software [30]. Just as MoRAine 1.0, the second release is an easy-to-use alternative for the computation of sequence logos and adjusted transcription factor binding sites, but it now provides increased accuracy at decreased running times and a simplified user-interface with fewer parameters to adjust.

Increased information content improvement with MoRAine 2.0

Figure 6.1 illustrates the output of the MoRAine online service for the binding sites of the transcription factor RamB of *Corynebacterium glutamicum*. The TFBMs have been taken from CoryneRegNet release 5.0. As in most databases, in CoryneRegNet [12], each binding site is annotated in $5' \rightarrow 3'$ direction relative to the regulated target gene. By using MoRAine 2.0 the average information content is improved from 0.64 (original database TFBMs) to 1.15 (MoRAine-adjusted TFBMs) by switching the strands for 15 of the 38 input sequences. The computation time was less than 2 seconds.

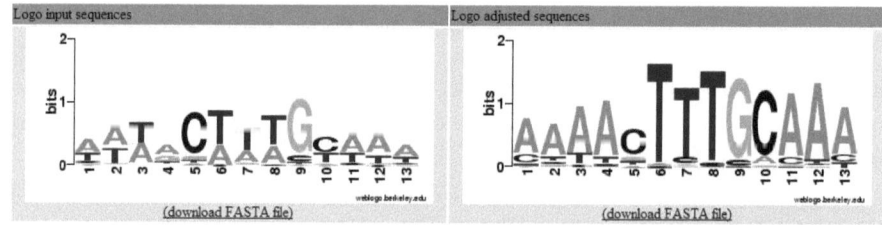

Fig. 6.1: A screenshot from the MoRAine 2.0 web site. A comparison of the sequence logos constructed from the original TFBMs (left side) for the transcription factor RamB of *Corynebacterium glutamicum* and the adjusted TFBMs by using MoRAine 2.0 (right side).

To demonstrate the performance, i.e. decreased running time and increased information content improvement, of MoRAine 2.0, the same datasets as in [16] were used: 1165 binding sites of 85 transcription factors of *Escherichia coli*. The average runtime and the mean information content improvement of MoRAine 2.0 are compared to the four methods implemented in MoRAine 1.0 for different lengths of the flanking sequences (l and r, respectively). As shown in Table 6.1, with MoRAine 1.0 the combination $(cg/simC)$ had the best runtime, but to gain the best information content improvement, one would use the combination $(cg/simS)$ [16]. With MoRAine 2.0, the gap between running time and accuracy has been closed. In Table 6.2, MoRAine 2.0 is compared with release 1.0 using the most accurate combination $(cg/simS)$ and the fastest combination $(cg/simC)$, respectively. For a fair running time comparison, MoRAine 1.0 $(cg/simC)$ and MoRAine 2.0 are re-evaluated on the same standard desktop PC. Table 6.2 shows that MoRAine 2.0 outperforms the previous release in terms of information content improvement with running times almost as fast as those of $(cg/simC)$. Furthermore, MoRAine 2.0 does not require the user to choose various input parameters to optimize its results.

6.1. MORAINE

	Difference (%)			
$l=r$	$cg/simC$	$cg/simS$	$km/simC$	$km/simS$
0	26.1	**27.0**	26.5	26.8
1	50.9	**54.4**	50.1	52.3
2	57.5	**63.6**	57.6	62.4
3	60.0	**69.5**	64.6	64.7
4	65.3	**70.1**	65.0	69.3
5	66.3	73.0	68.8	**73.3**
6	66.6	73.1	74.3	**74.9**
7	68.0	**78.7**	73.5	78.4
	Time (s)			
$l=r$	$cg/simC$	$cg/simS$	$km/simC$	$km/simS$
0	**0.6**	0.7	1.2	1.1
1	**0.7**	2.3	7.2	4.0
2	**0.8**	4.2	45.9	8.3
3	**1.0**	8.4	128.0	12.8
4	**1.1**	11.9	198.3	19.5
5	**1.3**	16.8	298.3	30.5
6	**1.8**	23.9	427.0	34.4
7	**2.0**	30.1	505.4	42.6

Tab. 6.1: This table was taken from [16] and summarizes the average information content improvements and the mean running times of MoRAine 1.0 for different l- and r-values and the four search method/similarity function combinations over all TFBMs of 85 transcriptional regulators of *E. coli*.

	Difference (%)		Time (s)	
$l=r$	MoRAine 1.0 ($cg/simS$)	MoRAine 2.0	MoRAine 1.0 ($cg/simC$)	MoRAine 2.0
0	27.0	27.2	0.21	0.23
1	54.4	54.7	0.26	0.29
2	63.6	66.5	0.32	0.36
3	69.5	72.2	0.38	0.42
4	70.1	75.5	0.46	0.50
5	73.0	75.7	0.55	0.59
6	73.1	77.8	0.60	0.66
7	78.7	79.1	0.71	0.77

Tab. 6.2: This table shows a comparison between the average information content improvement and the mean running time of MoRAine 1.0 and MoRAine 2.0 for different l- and r-values over all TFBMs of 85 transcriptional regulators of *E. coli*. MoRAine 2.0 is compared with the most accurate combination of similarity function and search method of MoRAine 1.0 (left side) and with the fastest combination (right side).

Predicting binding sites with adjusted TFBMs

The PSSMs derived from TFBMs are often used to predict further TFBMs in a given set of DNA sequences, generally in sequences upstream of putatively regulated target genes or operons. A PSSM allows to assign a score to any length-m DNA sequence window. We say that a PSSM matches such a window if the score exceeds a given threshold. A match is considered to be a good candidate for a real TFBM if the score is properly chosen (generally as the log-odds score between the nucleotide distribution of true binding sites on the one hand and a background distribution on the other hand) and the threshold (ideally based on statistical considerations; see e.g. [64]). Different algorithms and implementations exist to perform these searches. Here the tool PoSSuMsearch [17] is used for further analyses. It uses lookahead scoring and it is based on efficiently searching an enhanced suffix array that previously has been created from upstream sequences of *E. coli*. The threshold for a match is automatically computed based on the tolerable frequency of hits in random sequences (p-value) by an efficient and exact lazy-evaluation method (for more details refer to [17]).

In the following, PoSSuMsearch is used to evaluate the prediction performance of (1) PSSMs constructed from the original TFBMs extracted from the RegulonDB database and (2) the MoRAine-adjusted PSSMs using TC as clustering method. One will see that by using MoRAine for pre-processing, the classification performance is significantly increased. As mentioned before, 1165 TFBMs for 85 TFs from RegulonDB are extracted to construct 85 PSSMs. Additionally, 3341 upstream sequences of all transcriptional units of *E. coli* are obtained from CoryneRegNet (see Section 6.2). In CoryneRegNet, an upstream region is defined as that DNA sequence -560 to $+20$ bps upstream to the start codon of a transcriptional unit (a gene, or an operon respectively).

For each PSSM, both forward and reverse strand of upstream sequences are used to predict TFBMs with PoSSuMsearch, using different p-value thresholds. For each threshold, the precision, recall and F-measure are calculated, where

FP number of incorrectly predicted motifs

FN number of wrongly not predicted motifs

TP number of correctly predicted motifs

In Figure 6.2, precision vs. recall are plotted for varying p-value thresholds for all PSSMs readjusted with MoRAine 2.0 for $l = r = 4$ (blue curve) in comparison to the prediction performance obtained with the original PSSMs (red curve) and the performance using MoRAine 1.0 adjusted PSSMs (green curve). For a fixed recall, the MoRAine-adjusted precision is always higher than with the original, unadjusted TFBMs/PSSMs. Figure 6.3 plots the F-measure against different p-value thesholds. The plot show that predictions based on adjusted PSSMs outperform those based on original PSSMs for all thresholds.

6.2 CoryneRegNet

As shown in Chapter 5, TC is capable of providing meaningful information about protein clusters. Thus, TransClust has been integrated into CoryneRegNet. This section will describe how this integration was performed. Furthermore, a recently published application [12] where protein clusters obtained with TC are used to transfer gene regulatory networks from one model organism to closely related species, is presented.

6.2. CORYNEREGNET

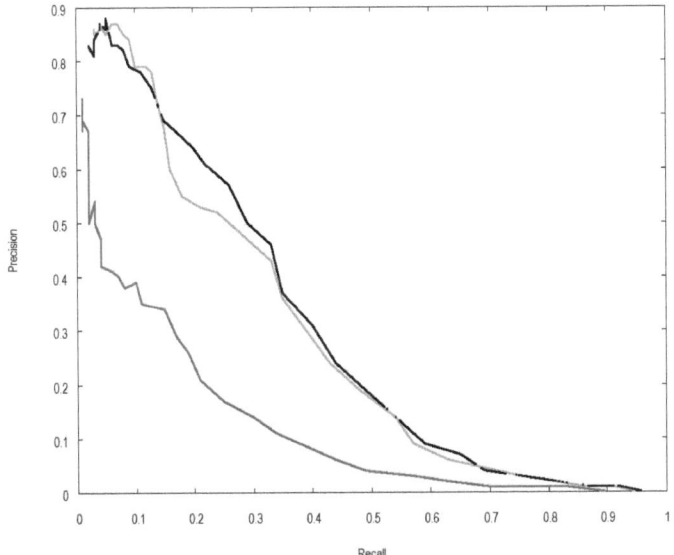

Fig. 6.2: Prediction performance comparison of PoSSuMsearch by means of precision and recall. All values are measured for varying p-value thresholds based on PSSMs learned from the original TFBMs (red line) compared to those of readjusted TFBMs with MoRAine 2.0 (blue line) and readjusted TFBMs with MoRAine 1.0 (green line).

CoryneRegNet [8] (online available at http://www.coryneregnet.de) is a reference database and analysis platform for Corynebacteria. It allows a pertinent data management of regulatory interactions along with the genome-scale reconstruction of transcriptional regulatory networks of corynebacteria relevant in human medicine and biotechnology, together with *Escherichia coli*. CoryneRegNet is based on a multi-layered, hierarchical and modular concept of transcriptional regulation and was implemented with an ontology-based data structure. It integrates the fast and statistically sound method PoSSuMsearch [17] to predict transcription factor binding sites within and across species. Reconstructed regulatory networks can be visualized on a web interface and as graphs. Special graph layout algorithms have been developed to facilitate the comparison of gene regulatory networks across species and to assist biologists with the evaluation of predicted and graphically visualized networks in the context of experimental results. To extend the comparative features, adequate data on gene and protein clusters is needed. The integration of this information widens the scope of CoryneRegNet, and assists the user with the reconstruction of unknown regulatory interactions [8–10].

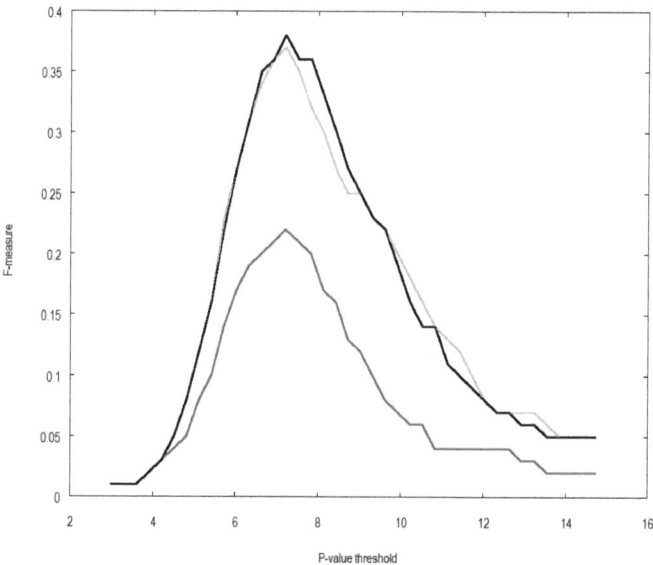

Fig. 6.3: Prediction performance comparison by means of plotting the F-measure for varying PoSSuMsearch p-value thresholds for the original TFBMs (red line), the MoRAine 1.0-adjusted TFBMs (green line), and the MoRAine 2.0-adjusted TFBMs (blue line) allowing 4 shifts to the left and right (l=r=4).

6.2.1 Integration of Transitivity Clustering with CoryneRegNet

Using TransClust, protein clusters for all organisms integrated in CoryneRegNet: *Corynebacterium diphtheriae, Corynebacterium efficiens, Corynebacterium glutamicum, Corynebacterium jeikeium, Escherichia coli, Mycobacterium tuberculosis CDC1551* and *Mycobacterium tuberculosis H37Rv* (altogether 22,797 proteins) are calculated. Based on the cluster size distribution, a comparatively high threshold of 30 was used with the similarity function SoH. This was empirically determined and can be explained by the relatively close evolutionary relationship of most organisms in CoryneRegNet.

The results computed by TransClust are parsed into the object oriented back-end and further on translated into the ontology based data structure of CoryneRegNet. The new concept class `ProteinCluster` has been added, as well as the relation type `bc` (belongs to cluster), which links the proteins to their clusters. Finally, the CoryneRegNet back-end has been adapted to import the new data into the database and the web-front-end to present the clusters.

6.2.2 Inter-species transfer of gene regulatory networks

In a recent study, Baumbach et al. [12] used protein clusters obtained by using TC to transfer gene regulatory networks from the model organism *Corynebacterium glutamicum* to the closely

6.2. CORYNEREGNET

related *Corynebacteria diphteria*, *C. jeikeium*, and *C. efficiens*. Together with the PSSM adjustment tool MoRAine and the binding site prediction software PoSSuMsearch a reliable prediction of whole networks was possible. Figure 6.4 illustrates the workflow. First all PSSMs for transcription factors of *C. glutamicum* were extracted from CoryneRegNet and improved using the MoRAine software. Afterwards PoSSuMsearch was applied to all upstream regions of genes in the target organisms to predict TFBMs. TC is used to identify clusters of homologous proteins. A regulation is considered to be conserved if (1) the source gene is conserved, (2) the binding site is conserved, and (3) the target gene is conserved as well. Choosing restrictive values for the parameters of the binding site prediction and the homology detection leads to a low number of false positives. In fact the parameters were adjusted to have no false positive prediction according to the known regulations in the database. By using this reliable network transfer, the database content was increased by a factor of 4.2. Table 6.3 summarizes the results for this study. In a recently published review by Venancio and Aravind [77], the workflow presented here was mentioned as the best current practice model.

	TFs	TFs$_C$	TFs^K		TFs$_C^K$		Regulations	
CG	128		69				530	
			original	transferred	transferred	original		transferred
CD	63	49 (77.8%)	2 (3.2%)	20 (× 10)	20 (40.1%)	46		193 (× 4.2)
CE	103	77 (74.8%)	5 (4.9%)	28 (× 5.6)	28 (36.4%)	64		348 (× 5.4)
CJ	55	31 (56.4%)	1 (1.8%)	13 (× 13)	13 (41.9%)	51		150 (× 2.9)
Av		69.6%	3.3%	× 9.5	39.7%			× 4.2

Tab. 6.3: Comparison between the original and the transferred database content of CoryneRegNet. Abbreviations: CG = *C. glutamicum*, CD = *C.diphteriae*, CE = *C. efficiens*, CJ = *C. jeikeium*, Av = Average, TFs = Transcription factors, TFs$_C$ = Common TFs with CG, TFs^K = TFs with known regulations, TFs$_C^K$ = Common TFs with CG that have known regulations. Taken from [12]

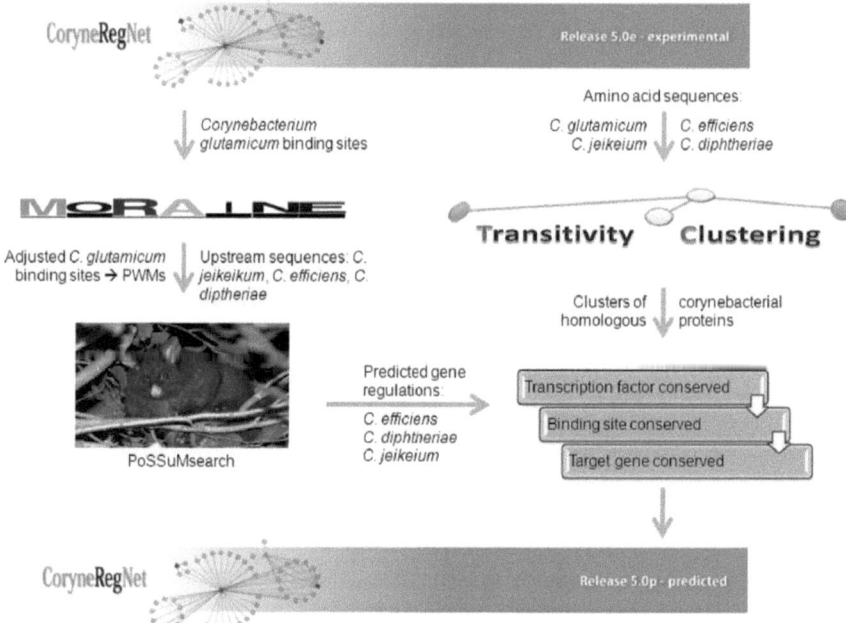

Fig. 6.4: Illustration of the workflow for gene regulatory network transfer. Starting with the database content of CoryneRegNet (1) the transcription factors and their known binding sites are extracted, (2) clusters of homologous proteins are calculated with FORCE, (3) the Position Specific Scoring Matrices are improved with MoRAine, and (4) afterwards used to predict binding sites in the upstream regions of all genes with PoSSuMsearch. Predicted regulations are entered in a new version of CoryneRegNet (5.0 predicted)

7. DISCUSSION

This work presented the clustering approach Transitivity Clustering. TC has proven to be useful in various bioinformatic tasks and even outperforms popular existing approaches like MCL or AP. Together with the software environment TransClust, TC contributes to all the important steps of a cluster analysis.

7.1 Transitivity Clustering and TransClust

In the introduction of this work (Chapter 1, page 7), a list of crucial and highly valuable necessities for a successful cluster analysis were discussed. As demonstrated in Chapter 2, many approaches only partially address these criteria. In the following, it will be discussed how TC and its implementation, TransClust, contribute to each point.

Similarity function: TransClust implements several methods of calculating similarities between two sequences based on BLAST results. The Cytoscape plugins "BLAST2SimGraph" and "ClusterExplorer", which are part of the TC Cytoscape plugin framework, integrate methods for computing and evaluating similarity measures. If a gold standard assignment for a related problem is given, or cluster assignments of a subset of the data is known, these may be utilized to investigate the appropriateness of the similarity measure.

Density parameter: TC needs only one single intuitive density parameter, a similarity threshold. The threshold defines what is considered as "similar enough". This value may be known in advance or can be guessed by a specialist in the field of the given application. It has been proven that a clustering obtained with TC follows certain rules that depend on this density parameter and the similarity function: The average similarity within a cluster is always above the threshold, while the average similarity between two clusters is always below the threshold. These conditions even partially hold for the extensions of TC, i.e. for overlapping or hierarchical clustering. Again, additional methods in the associated Cytoscape plugins have been implemented to ease the detection of a meaningful density parameter; see the example in Section 5.2.3 and Figure 5.5 on page 80.

Runtime and space efficiency: Although the graph modification problem underlying TC is NP-complete and even APX-hard, the TC approach and its efficient implementation, TransClust, have proven to be applicable even for large scale experiments. Clustering of more than 100,000 objects is possible in only a few minutes on a standard desktop PC. This is, on the one hand, due to the efficient implementation of TransClust, which combines very fast and accurate heuristics with an exact method. On the other hand, the WTGPP may be solved separately for the connected components of the input graph. Often, initially huge problems can be split into multiple smaller problems, which can be solved within a drastically reduced running time. This property is particularly beneficial if very sparse data serves as input.

Robustness against outliers and noise: As demonstrated in Chapter 5, TC is very robust against noise. The underlying WTGPP is further robust against outliers, since an assignment of a single element depends on its similarities to all other elements and their similarity to each other. Apparently, this is the reason why the transitivity concept is perfectly suited for clustering.

Interpretable results: Due to the intuitive density parameter, which directly "works" on the similarity function, the clustering output of TC has some intuitive properties that help to get an intuition about the final cluster assignment.

Integration of existing knowledge: Thanks to the flexibility of TC, the integration of existing knowledge is a straightforward task. Different methods have been implemented in TransClust: It is possible to set upper and lower bounds, to include an existing pre-clustering for a subset of the data. Accordingly, it is possible to directly specify whether two elements must be assigned to the same or to different clusters. A brief evaluation in Section 5.2.4 demonstrated the advantages. Little background information may already improve the quality of the clustering drastically.

Integration with standard software: TC has been successfully integrated with several other software systems: (1) the network analysis and visualization software Cytoscape, (2) the TFBM re-adjustment tool MoRAine, and (3) as homology prediction in the corynebacterial reference database CoryneRegNet.

Visualization: In order to not reinvent the wheel, and to profit from existing and specialized software, TransClust has been integrated into Cytoscape. The TransClust plugin framework for Cytoscape provides various options to import, pre-process, cluster, and visualize the data at each step of the typical data analysis workflow with just a few mouse-clicks.

Evaluation methods: The TransClust software framework provides the end-user with extensive aid with the following data analysis steps: (1) selection and configuration of an appropriate case-specific similarity measure, (2) estimation of a reasonable density parameter, and (3) evaluation and comparison of the clustering results. Note that the Cytoscape plugins also cover these tasks, but further allow answering to typical follow-up questions, such as the identification of outliers.

Reproducible results: In theory, the existence of more than one solution for a given WTGPP is possible. In practice, this almost never happens due to the real-valued similarity function.

7.2. COMPUTATIONAL BIOLOGY APPLICATIONS

On the implementation side, TransClust does not include any randomized algorithm but the parameter training, which is optional.

Small number of user-defined parameters: TC needs only one parameter, the density parameter. The TransClust framework asks for more parameters to control the trade-off between runtime and quality. These parameters are already optimized and usually do not need to be changed.

7.2 Computational biology applications

In addition to the theoretical attributes of TC, the applicability to real-life problems has been demonstrated by means of three case studies. The TC approach was compared to other commonly used clustering tools.

First, in Chapter 5, it has been demonstrated that TC is capable of clustering protein domains into superfamilies based on their sequence similarity alone; a long-standing challenge in computational biology. The corresponding analysis was performed similarly to the one by Paccanaro *et al.* [61]. A subset of the SCOP database served as gold standard reference here. TC performed best in comparison to Spectral Clustering, Markov Clustering, Hierarchical Clustering, GeneRAGE, and Affinity Propagation. Note that TC even outperformed the method of Paccanaro *et al.* (Spectral Clustering (SC)) on their own test data set and their quality measure.

In another study, whole protein sequences were clustered into families. A manually curated gold standard data set by Brown *et al.* [22] served as ground truth. Two other methods, namely MCL and AP were evaluated on the same data. Again, TC outperformed the other approaches. Furthermore, this study demonstrated the high influence of the density parameter on the resulting clusters. Consequently, the benefit of the threshold determination routine of TransClust has been showcased with this data set. However, finding a reasonable case-specific density parameter still remains problematic; even with background knowledge. Even worse, the limited amount of available, manually curated gold standard data so far hinders furthers analyses in this research area.

Last, we concentrated on the prediction of protein complexes in a given PPI network. Again, the evaluation was based on a previously published data. In 2006, Brohée *et al.* [21] compared four different clustering methods for this task and claimed MCL to be the best current practice method. The same data set and quality evaluation methods as in the original study were used to evaluate the applicability of TC. It revealed that TC can compete with MCL in terms of quality and outperforms it in terms of robustness. Since protein complexes do not need to be disjoint, i.e. one protein may contribute to multiple complexes, these data sets have been ideal to demonstrate the power of the overlapping clustering methods of TC.

To sum it all up, for all biological application cases with hand-curated data, TC competes or even outperforms other state-of-the-art clustering tools in terms of accuracy and robustness.

7.3 Integration in bioinformatics tools

Inter-species gene regulatory network transfer As demonstrated above, TC is suited to detect clusters of homologous proteins. In a recently published bioinformatics pipeline, these predictions were utilized to transfer gene regulatory networks from a model organism to taxonomically closely related species. This method has recently been cited as best current practice model.

Binding motif re-adjustment TC is part of the MoRAine software, a tool to re-adjust TFBMs by utilizing clustering methods. The initial release MoRAine 1.0 implemented clustering methods, which were either accurate or fast but not both. By using TC, this gap between accuracy and running time was closed. MoRAine 2.0 now produces more accurate results in almost the same time as the previous release.

Cytoscape The TC plugin framework for Cytoscape allows for an extended visual, graph-based data analysis. Cytoscape provides an end-user with various possibilities to import, layout, and analyze data that can be represented as graphs. Aside from TC, plugins for MCL, hierarchical clustering, and MCODE have been developed. But the TC plugin framework for Cytoscape offers more than just a clustering method: In contrast to all other tools it integrates various methods to help in each step of a typical cluster analysis.

7.4 Future directions

Although TC can be applied for various tasks, it has its limitations. This section describes ideas to overcome these problems together with more general future directions.

Huge data sets Like many other clustering tools, the applicability of TC is limited with the problem size. For very dense similarity graphs, it may be impossible to split the graph into multiple connected components. Such instances can still be clustered with TC but it requires more time since the connected components are rather large. Modifications to the heuristics to decrease the runtime may be inevitable; but may lead to reduced accuracy. To cluster billions of data objects, further modifications are necessary. Most obviously, in practice it is impossible for any clustering software to store all similarities for each pair of objects for such huge data sets. A putative solution would be the on-the-fly calculation of the similarities while clustering the data, i.e. iterate through the list of elements and decide in each step if it fits to one of the afore processed elements. To decide whether an element fits to one cluster, one needs to calculate the similarities to all already processed objects above a certain threshold. In such a way, connected components can be calculated even for very large data sets. These components are much smaller and can then serve as input for a more precise clustering approach like TC.

Fixed number of clusters One advantage of TC is its flexibility. However, there still exist limitations in the applicability of TC. If, for instance, a specific number of clusters is required

7.4. FUTURE DIRECTIONS

or the problem aims to find central elements, a different clustering approach might be better suited. One idea for the application of TC on such tasks is to add a constraint to the WTGPP that allows only a fixed number of clusters. This problem is known as MinDisAgree[k], where k is the number of clusters. It would be interesting to investigate how existing algorithms to solve the WTGPP may be extended to the MinDisAgree[k] problem.

Evaluation problems Most of all, the last evaluation study (PPI network clustering) points out the importance of the utilized quality measure; the results may differ strongly. The proposed quality measures of the original study clearly favored MCL. By choosing the F-measure, this conclusion could not be validated. Equally important is the choice of the gold standard. One example: Brohée et al. compared different methods on real-world data against a gold standard; the problem, however, was that this gold standard data only partially intersects with the real-world data sets that have been used for the evaluation by Brohée et al. In consequence, not only hand-curated assignments influenced the resulting ranking of the compared tools. Generally, this shows how careful an evaluation has to be performed. Many factors have to be taken into account, which are: the choice of the data and the gold standard assignment, the algorithm parameters for a fair comparison, and the quality measures. For future evaluations it is desired to have much larger, high-quality gold standard data sets and a standardized evaluation method. The work of Tompa *et al.* [74] about the standardized analysis of *de novo* motif discovery tools may serve as an example here.

In summary, future directions are the continuous improvement and development of further heuristics for the WTGPP and new areas of application with standardized evaluation routines.

8. CONCLUSION

This thesis introduced Transitivity Clustering, a clustering method based on weighted transitive graph projection, which aims for unraveling hidden transitive substructures in a given similarity graph deduced from a pairwise similarity measure.

TC attacks and solves all the crucial problems of a typical cluster analysis. The results of TC have, for instance, provable attributes that depend on the similarity function and the density parameter. This eases the detection of a meaningful, application-specific density parameter and aids with the results interpretation. Additionally, TC allows for a direct inclusion of background knowledge into the clustering process to further improve the accuracy. Another feature, which distinguishes TC from other clustering approaches, is its capability of producing partitional, overlapping and hierarchical clusterings. One may now use one model for each of these three different kinds of clustering and hence avoid using completely different approaches, whose results are hardly comparable.

The applicability of TC on bioinformatics tasks has been demonstrated by means of solving real-world clustering problems with hand-curated biological data sets. The identification of protein families and superfamilies based on their sequence similarity is only one example where TC outperforms existing standard clustering tools, such as Markov Clustering and Affinity Propagation.

TransClust is the efficient implementation of TC. It offers various methods to aid with each step of a cluster analysis. The standalone version allows for clustering of large data sets. Plugins for Cytoscape allow for an extended visual analysis and data investigation. TransClust received external interest and has been integrated with other bioinformatics applications. In the corynebacterial reference database CoryneRegNet, TransClust applies TC to predict groups of homologous proteins. As part of an internationally recognized computational pipeline, these predictions were utilized for the inter-species transfer of gene regulatory networks. Furthermore, the transcription factor binding motif re-adjustment tool MoRAine was improved by integrating TC.

Hence, TC is a novel, comprehensive clustering framework with special focus on applications in computational biology.

BIBLIOGRAPHY

[1] D. Aloise, A. Deshpande, P. Hansen, and P. Popat. NP-hardness of Euclidean sum-of-squares clustering. *Machine Learning*, 75:245–248, 2009.

[2] S. F. Altschul, T. L. Madden, A. A. Schäffer, J. Zhang, Z. Zhang, W. Miller, and D. J. Lipman. Gapped BLAST and PSI-BLAST: a new generation of protein database search programs. *Nucleic Acids Research*, 25(17):3389–3402, Sep 1997.

[3] A. Andreeva, D. Howorth, S. E. Brenner, T. J. Hubbard, C. Chothia, and A. G. Murzin. SCOP database in 2004: refinements integrate structure and sequence family data. *Nucleic Acids Research*, 32:D226–D229, 2004.

[4] B. Andreopoulos, A. An, X. Wang, and M. Schroeder. A roadmap of clustering algorithms: finding a match for a biomedical application. *Briefings in Bioinformatics*, Feb 2009.

[5] M. M. Babu, N. M. Luscombe, L. Aravind, M. Gerstein, and S. A. Teichmann. Structure and evolution of transcriptional regulatory networks. *Current Opinion in Structural Biology*, 14(3):283–291, Jun 2004.

[6] M. M. Babu, S. A. Teichmann, and L. Aravind. Evolutionary dynamics of prokaryotic transcriptional regulatory networks. *Journal of Molecular Biology*, 358(2):614–633, Apr 2006.

[7] G. D. Bader and C. W. V. Hogue. An automated method for finding molecular complexes in large protein interaction networks. *BMC Bioinformatics*, 4:2, Jan 2003.

[8] J. Baumbach. CoryneRegNet 4.0 - A reference database for corynebacterial gene regulatory networks. *BMC Bioinformatics*, 8(1):429, Nov 2007.

[9] J. Baumbach and L. Apeltsin. Linking Cytoscape and the corynebacterial reference database CoryneRegNet. *BMC Genomics*, 9:184, 2008.

[10] J. Baumbach, K. Brinkrolf, L. F. Czaja, S. Rahmann, and A. Tauch. CoryneRegNet: an ontology-based data warehouse of corynebacterial transcription factors and regulatory networks. *BMC Genomics*, 7:24, 2006.

[11] J. Baumbach, K. Brinkrolf, T. Wittkop, A. Tauch, and S. Rahmann. CoryneRegNet 2: An integrative bioinformatics approach for reconstruction and comparison of transcriptional regulatory networks in prokaryotes. *Journal of Integrative Bioinformatics*, 3(2):24, 2006.

[12] J. Baumbach, S. Rahmann, and A. Tauch. Reliable transfer of transcriptional gene regulatory networks between taxonomically related organisms. *BMC Systems Biology*, 3:8, 2009.

[13] J. Baumbach, A. Tauch, and S. Rahmann. Towards the integrated analysis, visualization and reconstruction of microbial gene regulatory networks. *Briefings in Bioinformatics*, 10(1):75–83, Jan 2009.

[14] J. Baumbach, T. Wittkop, C. K. Kleindt, and A. Tauch. Integrated analysis and reconstruction of microbial transcriptional gene regulatory networks using CoryneRegNet. *Nature Protocols*, 4(6):992–1005, 2009.

[15] J. Baumbach, T. Wittkop, K. Rademacher, S. Rahmann, K. Brinkrolf, and A. Tauch. CoryneRegNet 3.0–an interactive systems biology platform for the analysis of gene regulatory networks in corynebacteria and *Escherichia coli*. *Journal of Biotechnology*, 129(2):279–289, Apr 2007.

[16] J. Baumbach, T. Wittkop, J. Weile, T. Kohl, and S. Rahmann. MoRAine - A web server for fast computational transcription factor binding motif reannotation. *Journal of Integrative Bioinformatics*, 5(2):91, 2008.

[17] M. Beckstette, R. Homann, R. Giegerich, and S. Kurtz. Fast index based algorithms and software for matching position specific scoring matrices. *BMC Bioinformatics*, 7:389, 2006.

[18] A. Ben-Dor, R. Shamir, and Z. Yakhini. Clustering gene expression patterns. *Journal of Computational Biology*, 6(3-4):281–297, 1999.

[19] M. Blatt, S. Wiseman, and E. Domany. Superparamagnetic clustering of data. *Physical Review Letters*, 76(18):3251–3254, Apr 1996.

[20] B.-J. Breitkreutz, C. Stark, and M. Tyers. The GRID: the general repository for interaction datasets. *Genome Biology*, 4(3):R23, 2003.

[21] S. Brohée and J. van Helden. Evaluation of clustering algorithms for protein-protein interaction networks. *BMC Bioinformatics*, 7:488, 2006.

[22] S. D. Brown, J. A. Gerlt, J. L. Seffernick, and P. C. Babbitt. A gold standard set of mechanistically diverse enzyme superfamilies. *Genome Biology*, 7(1):R8, 2006.

[23] S. Böcker, S. Briesemeister, Q. B. A. Bui, and A. Truss. Going weighted: Parameterized algorithms for cluster editing. *Theoretical Computer Science*, 2009. doi:10.1016/j.tcs.2009.05.006.

[24] S. Böcker, S. Briesemeister, and G. W. Klau. Exact algorithms for cluster editing: Evaluation and experiments. *Algorithmica*, 2009.

[25] M. Cameron, Y. Bernstein, and H. E. Williams. Clustered sequence representation for fast homology search. *Journal of Computational Biology*, 14(5):594–614, Jun 2007.

[26] J.-M. Chandonia, G. Hon, N. S. Walker, L. L. Conte, P. Koehl, M. Levitt, and S. E. Brenner. The ASTRAL compendium in 2004. *Nucleic Acids Research*, 32:D189–D192, 2004.

[27] M. Charikar, V. Guruswami, and A. Wirth. Clustering with qualitative information. *Journal of Computer and System Sciences*, 71:360–383, 2003.

[28] D. S. Chekmenev, C. Haid, and A. E. Kel. P-Match: transcription factor binding site search by combining patterns and weight matrices. *Nucleic Acids Research*, 33(Web Server issue):W432–W437, Jul 2005.

[29] M. S. Cline, M. Smoot, E. Cerami, A. Kuchinsky, N. Landys, C. Workman, R. Christmas, I. Avila-Campilo, M. Creech, B. Gross, K. Hanspers, R. Isserlin, R. Kelley, S. Killcoyne, S. Lotia, S. Maere, J. Morris, K. Ono, V. Pavlovic, A. R. Pico, A. Vailaya, P.-L. Wang, A. Adler, B. R. Conklin, L. Hood, M. Kuiper, C. Sander, I. Schmulevich, B. Schwikowski, G. J. Warner, T. Ideker, and G. D. Bader. Integration of biological networks and gene expression data using Cytoscape. *Nature Protocols*, 2(10):2366–2382, 2007.

[30] G. E. Crooks, G. Hon, J.-M. Chandonia, and S. E. Brenner. WebLogo: a sequence logo generator. *Genome Research*, 14(6):1188–1190, Jun 2004.

[31] W. Duan, M. Song, and A. Yates. Fast max-margin clustering for unsupervised word sense disambiguation in biomedical texts. *BMC Bioinformatics*, 10 Suppl 3:S4, 2009.

[32] A. J. Enright, S. V. Dongen, and C. A. Ouzounis. An efficient algorithm for large-scale detection of protein families. *Nucleic Acids Research*, 30(7):1575–1584, Apr 2002.

[33] A. J. Enright and C. A. Ouzounis. GeneRAGE: a robust algorithm for sequence clustering and domain detection. *Bioinformatics*, 16(5):451–457, May 2000.

[34] B. J. Frey and D. Dueck. Clustering by passing messages between data points. *Science*, 315(5814):972–976, 2007.

[35] T. M. J. Fruchterman and E. M. Reingold. Graph drawing by force-directed placement. *Software - Practice and Experience*, 21(11):1129–1164, 1991.

[36] D. J. Galas and A. Schmitz. DNAse footprinting: a simple method for the detection of protein-DNA binding specificity. *Nucleic Acids Research*, 5(9):3157–3170, Sep 1978.

[37] A.-C. Gavin, P. Aloy, P. Grandi, R. Krause, M. Boesche, M. Marzioch, C. Rau, L. J. Jensen, S. Bastuck, B. Dümpelfeld, A. Edelmann, M.-A. Heurtier, V. Hoffman, C. Hoefert, K. Klein, M. Hudak, A.-M. Michon, M. Schelder, M. Schirle, M. Remor, T. Rudi, S. Hooper, A. Bauer, T. Bouwmeester, G. Casari, G. Drewes, G. Neubauer, J. M. Rick,

B. Kuster, P. Bork, R. B. Russell, and G. Superti-Furga. Proteome survey reveals modularity of the yeast cell machinery. *Nature*, 440(7084):631–636, Mar 2006.

[38] A.-C. Gavin, M. Bösche, R. Krause, P. Grandi, M. Marzioch, A. Bauer, J. Schultz, J. M. Rick, A.-M. Michon, C.-M. Cruciat, M. Remor, C. Höfert, M. Schelder, M. Brajenovic, H. Ruffner, A. Merino, K. Klein, M. Hudak, D. Dickson, T. Rudi, V. Gnau, A. Bauch, S. Bastuck, B. Huhse, C. Leutwein, M.-A. Heurtier, R. R. Copley, A. Edelmann, E. Querfurth, V. Rybin, G. Drewes, M. Raida, T. Bouwmeester, P. Bork, B. Seraphin, B. Kuster, G. Neubauer, and G. Superti-Furga. Functional organization of the yeast proteome by systematic analysis of protein complexes. *Nature*, 415(6868):141–147, Jan 2002.

[39] J. Gramm, J. Guo, F. Hüffner, and R. Niedermeier. Automated generation of search tree algorithms for hard graph modification problems. *Algorithmica*, 39(4):321–347, 2004.

[40] J. Gramm, J. Guo, F. Hüffner, and R. Niedermeier. Graph-modeled data clustering: Exact algorithm for clique generation. *Theoretical Computer Science*, 38(4):373–392, 2005.

[41] M. Grötschel and Y. Wakabayashi. A cutting plane algorithm for a clustering problem. *Mathematical Programming, Series B*, 45:59–96, 1989.

[42] J. A. Hartigan. *Clustering Algorithms*. Wiley, 1975.

[43] L. M. Hellman and M. G. Fried. Electrophoretic mobility shift assay (EMSA) for detecting protein-nucleic acid interactions. *Nature Protocols*, 2(8):1849–1861, 2007.

[44] Y. Ho, A. Gruhler, A. Heilbut, G. D. Bader, L. Moore, S.-L. Adams, A. Millar, P. Taylor, K. Bennett, K. Boutilier, L. Yang, C. Wolting, I. Donaldson, S. Schandorff, J. Shewnarane, M. Vo, J. Taggart, M. Goudreault, B. Muskat, C. Alfarano, D. Dewar, Z. Lin, K. Michalickova, A. R. Willems, H. Sassi, P. A. Nielsen, K. J. Rasmussen, J. R. Andersen, L. E. Johansen, L. H. Hansen, H. Jespersen, A. Podtelejnikov, E. Nielsen, J. Crawford, V. Poulsen, B. D. Sørensen, J. Matthiesen, R. C. Hendrickson, F. Gleeson, T. Pawson, M. F. Moran, D. Durocher, M. Mann, C. W. V. Hogue, D. Figeys, and M. Tyers. Systematic identification of protein complexes in *Saccharomyces cerevisiae* by mass spectrometry. *Nature*, 415(6868):180–183, Jan 2002.

[45] T. Ito, T. Chiba, R. Ozawa, M. Yoshida, M. Hattori, and Y. Sakaki. A comprehensive two-hybrid analysis to explore the yeast protein interactome. *Proceedings of the National Academy of Sciences of the United States of America*, 98(8):4569–4574, Apr 2001.

[46] S. J.Lange. Efficient weighted graph cluster editing using an enhanced layout-based approach. Master's thesis, Bielefeld University, 2008.

[47] R. Jothi, S. Cuddapah, A. Barski, K. Cui, and K. Zhao. Genome-wide identification of in vivo protein-DNA binding sites from ChIP-Seq data. *Nucleic Acids Research*, 36(16):5221–5231, Sep 2008.

[48] A. E. Kel, E. Gössling, I. Reuter, E. Cheremushkin, O. V. Kel-Margoulis, and E. Wingender. MATCH: A tool for searching transcription factor binding sites in DNA sequences. *Nucleic Acids Research*, 31(13):3576–3579, Jul 2003.

[49] A. D. King, N. Przulj, and I. Jurisica. Protein complex prediction via cost-based clustering. *Bioinformatics*, 20(17):3013–3020, 2004.

[50] N. Kleinboelting. Protein clustering using an ant colony layouting approach. Master's thesis, Bielefeld University, 2009.

[51] A. Krause, J. Stoye, and M. Vingron. Large scale hierarchical clustering of protein sequences. *BMC Bioinformatics*, 6:15, 2005.

[52] N. J. Krogan, G. Cagney, H. Yu, G. Zhong, X. Guo, A. Ignatchenko, J. Li, S. Pu, N. Datta, A. P. Tikuisis, T. Punna, J. M. Peregrín-Alvarez, M. Shales, X. Zhang, M. Davey, M. D. Robinson, A. Paccanaro, J. E. Bray, A. Sheung, B. Beattie, D. P. Richards, V. Canadien, A. Lalev, F. Mena, P. Wong, A. Starostine, M. M. Canete, J. Vlasblom, S. Wu, C. Orsi, S. R. Collins, S. Chandran, R. Haw, J. J. Rilstone, K. Gandi, N. J. Thompson, G. Musso, P. S. Onge, S. Ghanny, M. H. Y. Lam, G. Butland, A. M. Altaf-Ul, S. Kanaya, A. Shilatifard, E. O'Shea, J. S. Weissman, C. J. Ingles, T. R. Hughes, J. Parkinson, M. Gerstein, S. J. Wodak, A. Emili, and J. F. Greenblatt. Global landscape of protein complexes in the yeast *Saccharomyces cerevisiae*. *Nature*, 440(7084):637–643, Mar 2006.

[53] M. Křivánek and J. Morávek. NP-hard problems in hierarchical-tree clustering. *Acta Informatica*, 23(3):311–323, 1986.

[54] L. Li, C. J. Stoeckert, and D. S. Roos. OrthoMCL: identification of ortholog groups for eukaryotic genomes. *Genome Research*, 13(9):2178–2189, Sep 2003.

[55] S. Lloyd. Least squares quantization in PCM. *IEEE Transactions on Information Theory*, 28:129–137, 1982.

[56] L. Mao, J. L. V. Hemert, S. Dash, and J. A. Dickerson. Arabidopsis gene co-expression network and its functional modules. *BMC Bioinformatics*, 10:346, 2009.

[57] T. Meinel, A. Krause, H. Luz, M. Vingron, and E. Staub. The SYSTERS Protein Family Database in 2005. *Nucleic Acids Research*, 33(Database issue):D226–D229, Jan 2005.

[58] H. W. Mewes, C. Amid, R. Arnold, D. Frishman, U. Güldener, G. Mannhaupt, M. Münsterkötter, P. Pagel, N. Strack, V. Stümpflen, J. Warfsmann, and A. Ruepp. MIPS: analysis and annotation of proteins from whole genomes. *Nucleic Acids Research*, 32(Database issue):D41–D44, Jan 2004.

[59] R. Münch, K. Hiller, A. Grote, M. Scheer, J. Klein, M. Schobert, and D. Jahn. Virtual Footprint and PRODORIC: an integrative framework for regulon prediction in prokaryotes. *Bioinformatics*, 21(22):4187–4189, Nov 2005.

[60] C. O. Pabo and R. T. Sauer. Transcription factors: structural families and principles of DNA recognition. *Annual Review of Biochemistry*, 61:1053–1095, 1992.

[61] A. Paccanaro, J. A. Casbon, and M. A. Saqi. Spectral clustering of protein sequences. *Nucleic Acids Research*, 34(5):1571–1580, 2006.

[62] S. Philippi and J. Köhler. Addressing the problems with life-science databases for traditional uses and systems biology. *Nature Review Genetics*, 7(6):482–488, Jun 2006.

[63] V. J. Promponas, A. J. Enright, S. Tsoka, D. P. Kreil, C. Leroy, S. Hamodrakas, C. Sander, and C. A. Ouzounis. CAST: an iterative algorithm for the complexity analysis of sequence tracts. complexity analysis of sequence tracts. *Bioinformatics*, 16(10):915–922, Oct 2000.

[64] S. Rahmann, T. Mueller, and M. Vingron. On the power of profiles for transcription factor binding site detection. *Statistical Applications in Genetics and Molecular Biology*, 2(1), 2003.

[65] S. Rahmann, T. Wittkop, J. Baumbach, M. Martin, A. Truß, and S. Böcker. Exact and heuristic algorithms for weighted cluster editing. *Conference on Computational Systems Bioinformatics*, 6(1):391–401, Aug 2007.

[66] O. Resendis-Antonio, J. A. Freyre-González, R. Menchaca-Méndez, R. M. Gutiérrez-Ríos, A. Martínez-Antonio, C. Avila-Sánchez, and J. Collado-Vides. Modular analysis of the transcriptional regulatory network of *E. coli*. *Trends in Genetics*, 21(1):16–20, Jan 2005.

[67] E. W. Sayers, T. Barrett, D. A. Benson, E. Bolton, S. H. Bryant, K. Canese, V. Chetvernin, D. M. Church, M. Dicuccio, S. Federhen, M. Feolo, L. Y. Geer, W. Helmberg, Y. Kapustin, D. Landsman, D. J. Lipman, Z. Lu, T. L. Madden, T. Madej, D. R. Maglott, A. Marchler-Bauer, V. Miller, I. Mizrachi, J. Ostell, A. Panchenko, K. D. Pruitt, G. D. Schuler, E. Sequeira, S. T. Sherry, M. Shumway, K. Sirotkin, D. Slotta, A. Souvorov, G. Starchenko, T. A. Tatusova, L. Wagner, Y. Wang, W. J. Wilbur, E. Yaschenko, and J. Ye. Database resources of the National Center for Biotechnology Information. *Nucleic Acids Research*, Nov 2009.

[68] R. Shamir, R. Sharan, and D. Tsur. Cluster graph modification problems. *Discrete Applied Mathematics*, 144:173–182, 2004.

[69] P. Shannon, A. Markiel, O. Ozier, N. S. Baliga, J. T. Wang, D. Ramage, N. Amin, B. Schwikowski, and T. Ideker. Cytoscape: a software environment for integrated models of biomolecular interaction networks. *Genome Research*, 13(11):2498–2504, Nov 2003.

[70] T. F. Smith and M. S. Waterman. Identification of common molecular subsequences. *Journal of Molecular Biology*, 147(1):195–197, Mar 1981.

[71] N. Song, J. M. Joseph, G. B. Davis, and D. Durand. Sequence similarity network reveals common ancestry of multidomain proteins. *PLoS Computational Biology*, 4(4):e1000063, Apr 2008.

[72] G. D. Stormo. DNA binding sites: representation and discovery. *Bioinformatics*, 16(1):16–23, Jan 2000.

[73] L. V. Sun, L. Chen, F. Greil, N. Negre, T.-R. Li, G. Cavalli, H. Zhao, B. V. Steensel, and K. P. White. Protein-DNA interaction mapping using genomic tiling path microarrays in *Drosophila*. *Proceedings of the National Academy of Sciences of the United States of America*, 100(16):9428–9433, Aug 2003.

[74] M. Tompa, N. Li, T. L. Bailey, G. M. Church, B. D. Moor, E. Eskin, A. V. Favorov, M. C. Frith, Y. Fu, W. J. Kent, V. J. Makeev, A. A. Mironov, W. S. Noble, G. Pavesi, G. Pesole, M. Régnier, N. Simonis, S. Sinha, G. Thijs, J. van Helden, M. Vandenbogaert, Z. Weng, C. Workman, C. Ye, and Z. Zhu. Assessing computational tools for the discovery of transcription factor binding sites. *Nature Biotechnology*, 23(1):137–144, Jan 2005.

[75] P. Uetz, L. Giot, G. Cagney, T. A. Mansfield, R. S. Judson, J. R. Knight, D. Lockshon, V. Narayan, M. Srinivasan, P. Pochart, A. Qureshi-Emili, Y. Li, B. Godwin, D. Conover, T. Kalbfleisch, G. Vijayadamodar, M. Yang, M. Johnston, S. Fields, and J. M. Rothberg. A comprehensive analysis of protein-protein interactions in *Saccharomyces cerevisiae*. *Nature*, 403(6770):623–627, Feb 2000.

[76] S. van Dongen. *Graph Clustering by Flow Simulation*. PhD thesis, University of Utrecht,, 2000.

[77] T. M. Venancio and L. Aravind. Reconstructing prokaryotic transcriptional regulatory networks: lessons from actinobacteria. *Journal of Biology*, 8(3):29, 2009.

[78] J. Vlasblom and S. J. Wodak. Markov clustering versus affinity propagation for the partitioning of protein interaction graphs. *BMC Bioinformatics*, 10:99, 2009.

[79] T. Wittkop, J. Baumbach, F. Lobo, and S. Rahmann. Large scale clustering of protein sequences with FORCE – A layout based heuristic for weighted cluster editing. *BMC Bioinformatics*, 8(1):396, Oct 2007.

[80] C. T. Zahn Jr. Approximating symmetric relations by equivalence relations. *Journal of the Society of Industrial and Applied Mathematics*, 12(4):840–847, 1964.

ABBREVIATONS

ACC	Accuracy
AP	Affinity Propagation
BeH	Best Hit
BLAST	Basic Local Alignment Search Tool
CAST	Cluster Affinity Search Technique
Cov	Coverage
FN	False Negatives
FORCE	Force-Based Cluster Editing
FP	False Positives
FP	Fixed-Parameter
GUI	Graphical User Interface
HC	Hierarchical Clustering
HSP	High Scoring Pair
ILP	Integer Linear Programming
LP	Linear Programming
MCL	Markov Clustering
MCODE	Molecular Complex Detection
PFM	Position Frequency Matrix
PPI	Protein-Protein Interaction
PPV	Positive Predictive Value
PSSM	Position Specific Scoring Matrix
RNSC	Restricted Neighborhood Search Clustering

SC	Spectral Clustering
SPC	Super Paramagnetic Clustering
SoH	Sum of Hits
TC	Transitivity Clustering
TF	Transcription Factor
TFBM	Transcription Factor Binding Motif
TGP	Transitive Graph Projection
TGPP	Transitive Graph Projection Problem
TN	True Negatives
TP	True Positives
TU	Transcription Unit
WTGP	Weighted Transitive Graph Projection
WTGPP	Weighted Transitive Graph Projection Problem

LIST OF FIGURES

3.1	Illustration of the WTGPP.	31
3.2	Counter example, that the WTGPP with upper bound lead to higher costs.	35
3.3	Example, that the WTGP does not produce an hierarchical structure for ascending thresholds.	37
4.1	Illustration of the TransClust program structure.	48
4.2	Model of the data import process.	49
4.3	Model of the clustering process.	50
4.4	Illustration of the layout-based heuristic as it is implemented in TransClust.	51
4.5	Illustration of the force-based layout.	52
4.6	An overview of the recursive post-processing method	59
4.7	Illustration of the graphical user interface of TransClust	61
4.8	Illustration of the Cytoscape software with integrated TransClust plugin	63
4.9	Screenshot of the TransClust website.	64
4.10	Comparison of the Cluster Affinity Search Technique (CAST) heuristic and the force-based heuristic, using either no or the default recursive post-processing method.	67
4.11	Quality evaluation of the heuristic.	69
4.12	Runtime comparison between the exact fixed-parameter algorithm and TransClust.	70
4.13	Runtime comparison between the exact fixed-parameter algorithm and TransClust.	70
5.1	Illustrations of the results from the protein domain clustering comparsion	73
5.2	Graphical summary of the obtained clustering results of FORCE for protein domain clustering	75
5.3	Illustration of the impact of the choice of the density parameter on the quality of the resulting clustering.	77
5.4	Screenshot of the inter vs. intra edge weight distribution of the amidohydrolase superfamily using the ClusterExplorer Cytoscape plugin.	78
5.5	Illustration of a typical analysis work flow using TransClust	80
5.6	Illustration of the impact of the amount of additional knowledge added.	81
5.7	Illustration of the robustness analysis.	83
6.1	A screenshot from the MoRAine 2.0 web site.	92
6.2	Prediction performance comparison of PoSSuMsearch by means of precision and recall using original and MoRAine adjusted motifs.	95

6.3	Prediction performance comparison of PoSSuMsearch by means of the F-measure using original and MoRAine adjusted motifs.	96
6.4	Illustration of the workflow for gene regulatory network transfer.	98
D.1	Quality comparison between TransClust and FORCE on the COG data set. . . .	134
D.2	Runtime comparison between TransClust and FORCE on the COG data set. . .	135
D.3	Comparison of the different layout methods in TransClust.	136
D.4	Runtime comparison between fixed-parameter approach, FORCE, and Greedy heuristic.	137
D.5	Quality evaluation of FORCE and Greedy heuristic against exact solution obtained with fixed-parameter approach.	138
D.6	Protein-Protein interaction network that is used in the evaluations of Chapter 5	139
D.7	Results of the original robustness analysis of clustering algorithm by Brohée et al.	140
D.8	Illustration of the results of the robustness evaluation for TransClust. The used quality measure is the separation	141
D.9	Illustration of the results of the robustness evaluation for TransClust. The used quality measure is the accuracy	141

LIST OF TABLES

2.1	Comparison of different clustering approaches.	27
2.2	Overview of a variety of clustering algorithms and how they fulfill the desired features as specified in the requirement analysis in Chapter 1.	28
4.1	The dataset used for runtime and quality analysis of TransClust heuristics against fixed-parameter method.	65
4.2	Costs and time for clustering of the COG data set using different limitations for the exact Fixed-Parameter (FP) approach	66
4.3	Running times on protein similarity data after data reduction for fixed-parameter and integer linear programming approach.	68
5.1	Summary of the evaluation of protein domain clustering.	74
5.2	Summary of the results and parameters for FORCE for protein domain clustering	76
5.3	Summary of the used density parameters for the protein clustering evaluation.	77
5.4	List of optimal density parameters with corresponding F-measure for protein clustering.	77
5.5	Comparison between the different clustering approaches from Brohée et al. [21] including results for TransClust	84
5.6	Comparison between MCL, RNSC, and TC on large scale data.	86
5.7	Comparison of the overlapping methods of TC.	87
6.1	Quality and runtime of MoRAine 1.0	93
6.2	Quality and runtime comparison of MoRAine 1.0 and MoRAine 2.0	93
6.3	Comparison between the orignial and the transferred database content of CoryneReg-Net.	97
D.1	Comparison of layout dimensions on the COG data set.	133
D.2	Clustering with K-means and parameter training.	133

APPENDIX

ACKNOWLEDGMENTS

This work would not have been possible without the support of a lot of people.

I am particularly grateful for the help and the support of my supervisors Prof. Dr. Sven Rahmann and Prof. Dr. Jens Stoye. They helped me in various discussion and always gave me honest feedback. During the last years, I learned a lot because of their experience and their helpful advise.

Furthermore, I would like to express thanks to my co-workers, with whom I often discussed open problems. Here, I also like to thank Dr. Jan Baumbach. Furthermore, I like to thank my office mate Alexander Bunkowski and my colleagues Dorothea, Heiko, Jochen, Sebastian O., Sebastian B., Sita, Wiebke and all members of the Genome Informatics workgroup as well as the members of the graduate school.

For proofreading, I would like to thank Alexander, Alisa, and Jan.

I should not forget to thank my friends who supported me and ignored my absence during the last months of this thesis.

Furthermore, I would like to thank my family who supported me all my life and especially during my studies and my PhD studies. I hope they are as proud of me as I am of them.

Last but not least, I would like to thank my girlfriend Alisa, who has been always there for me, listened to my problems, and is my inspiration in many fields.

A. PUBLICATIONS & COOPERATIONS

Transitivity Clustering as clustering method using the WTGPP as model was published and presented at CSB 2009 [65]. The force-based heuristic, a greedy approximation, and a first version of the fixed-parameter approach were also included in this publication. This has been done in cooperation with Sebastian Böcker from Jena University, Marcel Martin from Dortmund University, and Jan Baumbach from the International Computer Science Institute (ICSI) in Berkeley. Following this, the software FORCE has been published in BMC Bioinformatics [79] together with Jan Baumbach and Francisco Lobo from Universidade Federal de Minas Gerais, Belo Horizonte, Brazil. The software MoRAine 1.0 was presented at the International Symposium on Integrative Bioinformatics (IB08) [16] as cooperative work with Jan Baumbach and Jochen Weile from Bielefeld University. The author of this thesis took part in the development of CoryneRegNet and the integration of TC. Corresponding publications are [11, 15], and recently also [14]. Furthermore, a publication summarizing the recent improvements to the TransClust framework is in preparation. It is expected to be published in 2010. In cooperation with Dorothea Emig from the Max-Planck Institute Saarbrücken, the Cytoscape plugins have been improved and will also be discussed in that publication. The integration of TC into the software MoRAine (MoRAine 2.0) has recently been submitted to the International Symposium on Integrative Bioinformatics (IB10).

A. PUBLICATIONS & COOPERATIONS

B. TRANSCLUST DATA FORMATS

This chapter describes the data formates that are used in the TransClust framework.

Costmatrix file Lemma 3.4 states that it is sufficient to solve the WTGPP of the connected components only, instead of using the whole graph. Thus all unnecessary information can be discarded to save space. Similarity values of nodes from different connected components are not necessary to save. Since the similarity function is required to be symmetric also only one direction has to be stored. TransClust supports the import of costmatrix files which correspond each to one connected components. These files start with the number of nodes in the first line, followed by the names of the objects in the connected component each in a separate line. The similarities are written below these information. Following the order as given above the objects are notated in the following as $o_1, ..., o_n$. The i-th line after the names contains the costs to remove or add the edge between o_{i-1} and $\{o_i, ..., o_n\}$ in a tab delimited format. Hereby the costs to remove an edge is a positive number, while the costs for adding an edge is negative. If the user has chosen to set an upper bound and thus merge all objects whose similarity exceeds this limit, the format slightly changes. The already produced costs for the merging operations, which are all costs for adding edges between objects of a set of object that is merged, are written in the second line after the size of the connected component. The names of these sets are stored, again tab delimited, in the subsequent lines, each set (which may consists of only one object) in one line. Finally, the costs between the sets are calculated and written as described above. All costmatrices are stored in a directory together with a file containing all already transitive components. This additional file is similar to the subsequently described results file. It is tab delimited, where the first column contains the names of the objects and the second column a different number for connected component/cluster. Either a directory of costmatrices and a transitive connected component file or a single costmatrix file may be chosen as input.

Similarity file A standard format to store similarity information is to use a tab delimited similarity file, where the first and second elements are objects and the third column contains the similarity values. For a large set of objects these files may become very big. To save memory space, TransClust accepts only files in a certain format. As usual the file should be tab delimited as described above, but all similarities for one object have to be written in consecutive lines. This allows TransClust to create an index and hence speed up the subsequent search for connected components given a certain threshold. The file can contain non symmetric similarities which are later made symmetric by TransClust by choosing the smaller value of the two directions. It is important that both directions are written in the file if the chosen similarity function is already symmetric, due to the fact that the import process of TransClust treats non existing similarities as zero and thus would set the similarity of this pair to zero as it is the smaller value. Such similarity file may be used to either create costmatrix files as described above or directly as input for clustering with varying thresholds (see Section 4.2.6 for a description of this method).

BLAST/FASTA file For the task of clustering sequences TransClust offers various methods to derive a pairwise similarity from a given BLAST and FASTA file. The FASTA file is needed as it contains the sequence length which is used to calculate the coverage of a HSP and may contain proteins which do not occur in the BLAST file, but should be included in the results. The BLAST file should be in the ten column format (using the -m 8 option for BLAST) to include every necessary information. From these files TransClust generates first a similarity file and with a given threshold the costmatrix files corresponding to the connected components.

Gold standard file TransClust integrates a method to compare the obtained clustering results to a gold standard. This can be used to either evaluate the clustering quality for one specific threshold or help finding the best density parameter for this problem (see Section 4.2.6 for details about this functionality). The corresponding file should be again tab delimited where the first column contains all names and the second column the corresponding gold standard cluster assignments. The cluster names can be arbitrary strings, whose only restriction is, that it should not contain any tab stops. It is important to use the same identifiers for the objects in the gold standard file and in the input for the clustering algorithm to guaranty a valid comparison.

Config file TransClust accepts also a config file, containing all available parameters of the program. It can be used to store the best configuration after a evolutionary parameter training (see Section 4.2.1) and to always use the same user defined configurations without specifying all parameters again for each run.

Info file Using TransClust as a commandline tool it is possible to create an info file, which summarizes the used parameters and information about the clustering. In contrast to the results file it does not include the clustering itself, but a list of all used connected components, the size of the corresponding clusters, the score, and how much time was needed to calculate the clustering. These information might be interesting for a comparison against other clustering tools and especially for an evaluation against other algorithms that solve the WTGPP.

Results file The results of a run with TransClust are stored in a tab delimited results file. This file can have two different formats, depending on the used method. Clustering data from costmatrices, independent of how they were created, produces a clustering file similar to the gold standard file. All objects which occur in either a costmatrix file, or the transitive connected component file are listed here in the first column together with its cluster in the second column.

Clustering data iteratively from a similarity file with a list of thresholds, leads to a results file that contains more information. Using a gold standard, the results file contains in each line the used threshold, the f-measure between the corresponding clustering and the gold standard, and the information about the element assignments. While these informations are separated by tabs, the clustering information divides each cluster by a semicolon and in each cluster the including elements by a comma. If no gold standard file is provided, the second column is just a dash, and the file may still be used to have information about the cluster size distribution for the different thresholds.

Known assignments file TransClust offers to import known assignments to improve the clustering quality. The corresponding file can have one of two possible formats. Known clusters can be imported by using a file similar to the gold standard file. The second option is to specify the relation between two objects directly. In a three column tab delimited file the corresponding information are stored similar to the similarity file. The first and second column contain the names of the object and the third column specifies whether these two objects should be in one cluster (1) or not (-1).

C. MORAINE 1.0

This section describes the similarity measures and clustering strategies that were used in the MoRAine 1.0 software; for missing definitions and more information refer to Section 6.1.

Similarity Measures

*Motif-cluster similarity (**simC**)* To measure the similarity between a single TFBM s and an existing non-empty cluster C', one calculates the mean information content I for the frequency matrix constructed from all TFBMs of C' and s itself. It will be subsequently referred to as the motif-cluster similarity $simC(s, C')$.

*Motif-seed similarity (**simS**)* Following another strategy, each cluster is represented by a single seed motif s'. Here one calculates I for the frequency matrix built from only the seed motif and the new TFBM s. This value is called the motif-seed similarity $simS(s, s')$; it is faster to evaluate, but less accurate than $simC$.

These definitions apply only if the cluster C' to which a new motif s from a set S_i is to be assigned does not yet contain another motif from S_i. Otherwise, the similarity is set to $-\infty$; this ensures that each cluster contains only one motif from every set S_i. Respectively the Hamming similarity between two motifs from the same set S_i is set to $-\infty$.

Clustering strategies

The goal is to partition the set of motifs into $M = 2 \cdot (l + r + 1)$ clusters, where each cluster contains exactly n motifs, one of each S_i ($i = 1, \ldots, n$) and thus is a putative solution. The clustering strategies are:

*Variant of **k**-means with random seeds (**km**)* In this application, the number M of clusters is known; so one can use a variation of the k-means algorithm [42]. In the end, the cluster with the highest mean information content I is chosen. Starting with a random set of M (out of $n \cdot M$) motifs (the *seeds*) that form the initial clusters, the following procedure is iterated until convergence: Each motif, in arbitrary but fixed order, is assigned to the cluster that maximizes the similarity ($simC$ or $simS$) value. This results in M clusters, each consisting of n motifs. A new seed sequence is chosen for each cluster as the sequence that best represents the cluster. This continues until no more changes occur for the seed sequence set; see Algorithm 3 for details. This strategy can be repeated for different initial seeds and addition orders.

Algorithm 3 k-means variant (km)
Input: sets S_i, $i = 1, \ldots, n$, with $|S_i| = M$; a similarity function sim
Output: Set C of motifs with high information content I
1: $oldseeds \leftarrow \{\}$
2: $seeds \leftarrow \{M \text{ arbitrary elements of } \bigcup_{i=1}^{n} S_i\}$
3: **while** $seeds \neq oldseeds$ **do**
4:      initialize clusters C_j, $j = 1, \ldots, M$, with one seed per cluster
5:      $oldseeds \leftarrow seeds$
6:      **for** $i \leftarrow 1$ to n **do**
7:          **for** all motifs s in S_i **do**
8:              assign s to cluster C_j with maximal $sim(s, C_j)$ over $j = 1, \ldots, M$
9:      $seeds \leftarrow \{\}$
10:     **for** all clusters C_j **do**
11:         find motif $s \in C_j$ with maximal $\sum_{s' \in C_j} sim(s, s')$
12:         add s to $seeds$
13: $C \leftarrow C_j$, with maximal $I(F_{C_j})$ over $j = 1, \ldots, M$
14: **return** $(C, I(F_C))$

Cluster growing (*cg*) Since each motif of each S_i must be in a different cluster, each S_i is used in turn as a set of initial seeds. Subsequently, the other motifs are added to their most similar cluster, similarly to the first iteration of the km algorithm, but this procedure is not iterated. Finally, the best solution obtained from the n different starting configurations is reported (see Algorithm 4 for details).

Algorithm 4 Cluster growing (*cg*)
Input: sets S_i, $i = 1, \ldots, n$, with $|S_i| = M$; a similarity function sim
Output: Set C of motifs with high information content I
1: $I_{best} \leftarrow 0$, $C_{best} \leftarrow \{\}$
2: **for** $i \leftarrow 1$ to n **do**
3:      $seeds \leftarrow S_i$
4:      initialize clusters C_j, $j = 1, \ldots, M$, with one seed per cluster
5:      **for** each $k \neq i$ **do**
6:          **for** all motifs s in S_k **do**
7:              assign s to C_j with maximal $sim(s, C_j)$ over $j = 1, \ldots, M$
8:      $C \leftarrow C_j$, with maximal $I(F_{C_j})$ over $j = 1, \ldots, M$
9:      **if** $I(F_C) \geq I_{best}$ **then**
10:         $I_{best} \leftarrow I$, $C_{best} \leftarrow C$
11: **return** (C_{best}, I_{best})

D. SUPPLEMENTARY FIGURES AND TABLES

Layout dimension	Cost	Time
2D	4,408,598	13 h 40 min 35 s
3D	4,413,944	12 h 58 min 33 s
4D	4,407,247	12 h 30 min 13 s
5D	4,406,788	13 h 14 min 36 s
6D	4,403,265	14 h 47 min 20 s
7D	4,410,258	15 h 01 min 06 s
8D	4,405,635	17 h 16 min 41 s

Tab. D.1: Comparison of layout dimensions on the COG data set. The force-based method is used together with single linkage clustering, recursive post-processing and parameter training for each problem instance. The table was taken from [46]

Layout dimension	Cost	Time
2D	4,425,695	13 h 55 min 14 s
3D	4,466,969	13 h 25 min 17 s
4D	4,440,647	13 h 59 min 58 s
5D	4,458,026	15 h 08 min 41 s
6D	4,433,953	16 h 32 min 26 s

Tab. D.2: Clustering results of the COG data set for different layout dimensions. The used methods were the force-based layout algorithm, K-means as geometric clustering, recursive post-processing, and parameter training for each instance. Taken from [46]

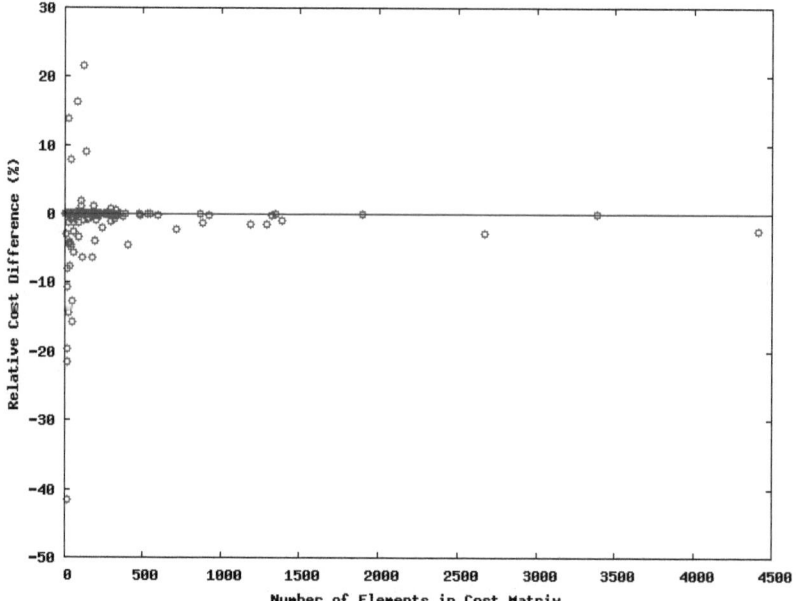

Fig. D.1: Quality comparison between TransClust and FORCE on the COG data set. Negative values correspond to instances where FORCE results in higher costs than TransClust, and positive results vice versa. Taken from [46]

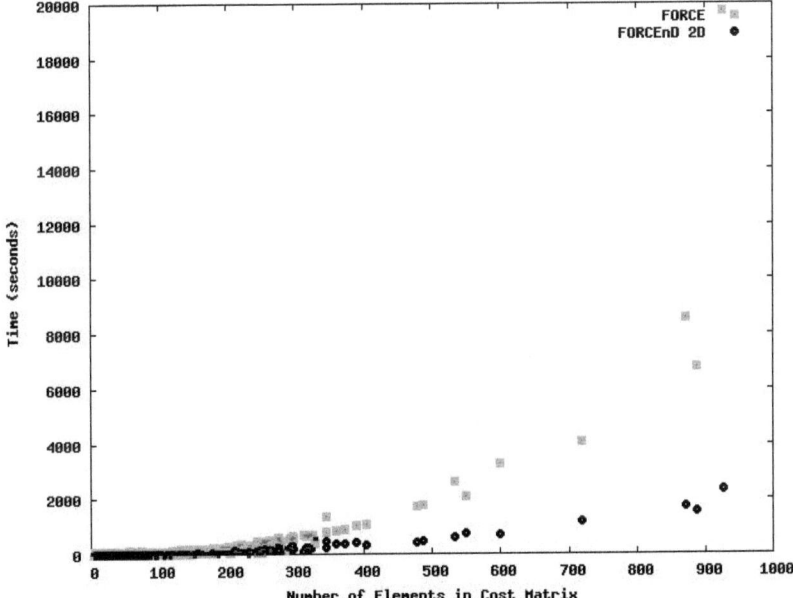

Fig. D.2: Runtime comparison between TransClust and FORCE on the COG data set up to 1000 nodes per problem instance. Taken from [46]

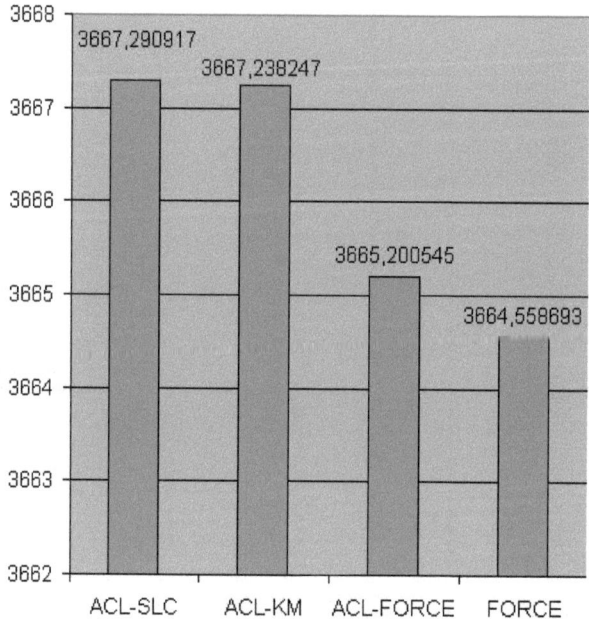

Fig. D.3: Comparison of the different layout methods in TransClust. The x-axis shows the costs of the different methods performed on an artificial data set. Abbreviations: ACL-SL: ant colony layout with single linkage as geometric clustering, ACL-KM: ant colony layout with K-means as geometric clustering, ACL-FORCE: ant colony layout as pre-processing for force-based layout with single linkage clustering, FORCE: force-based layout with single linkage clustering. Taken from [50]

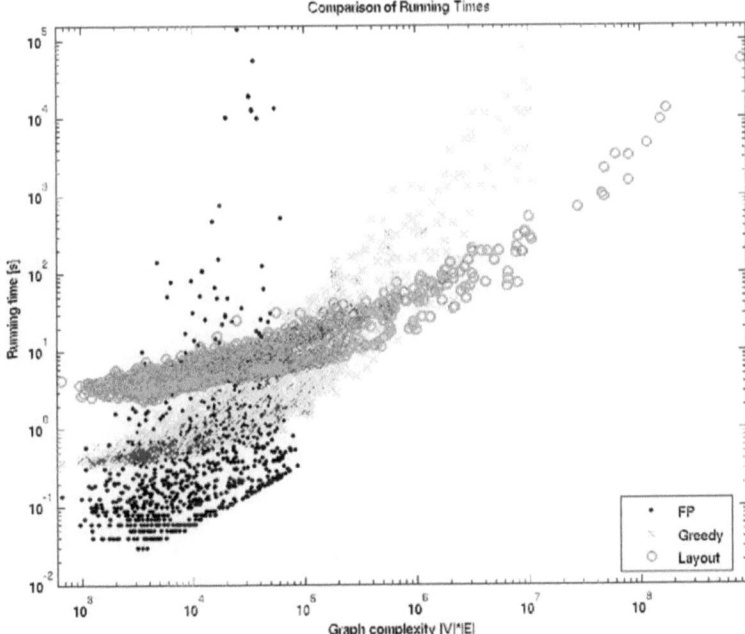

Fig. D.4: Runtime comparison between fixed-parameter approach, FORCE, and Greedy heuristic. Taken from [65]

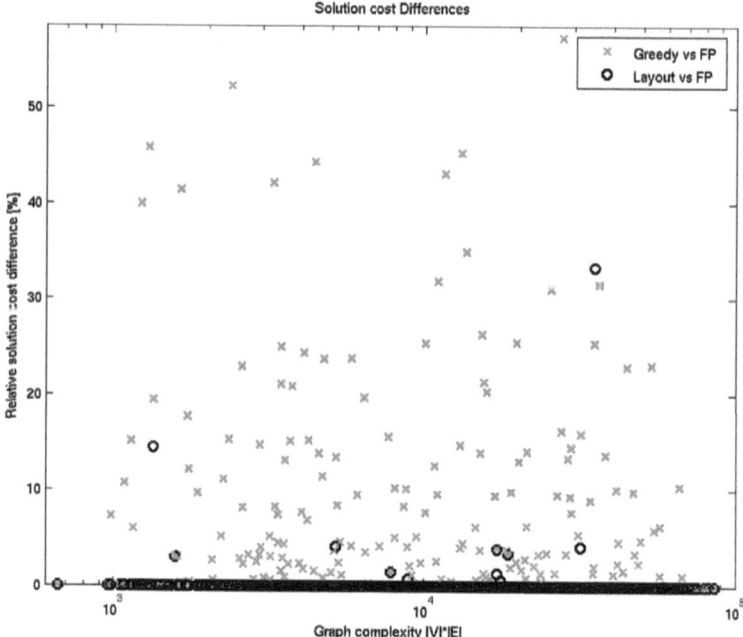

Fig. D.5: Quality evaluation of FORCE and Greedy heuristic against exact solution obtained with fixed-parameter approach. Taken from [65]

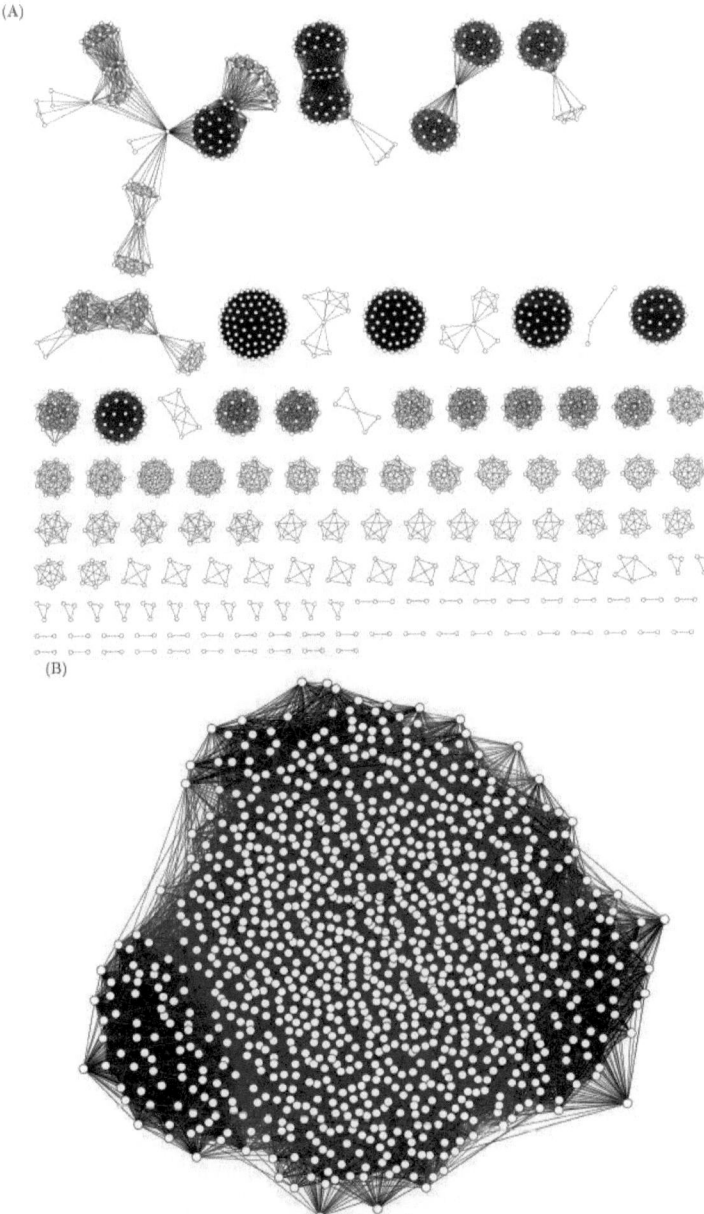

Fig. D.6: Graph of PPI network with (A) no edge modifications and (B) 100% added and 40% removed edges. Taken from [21]

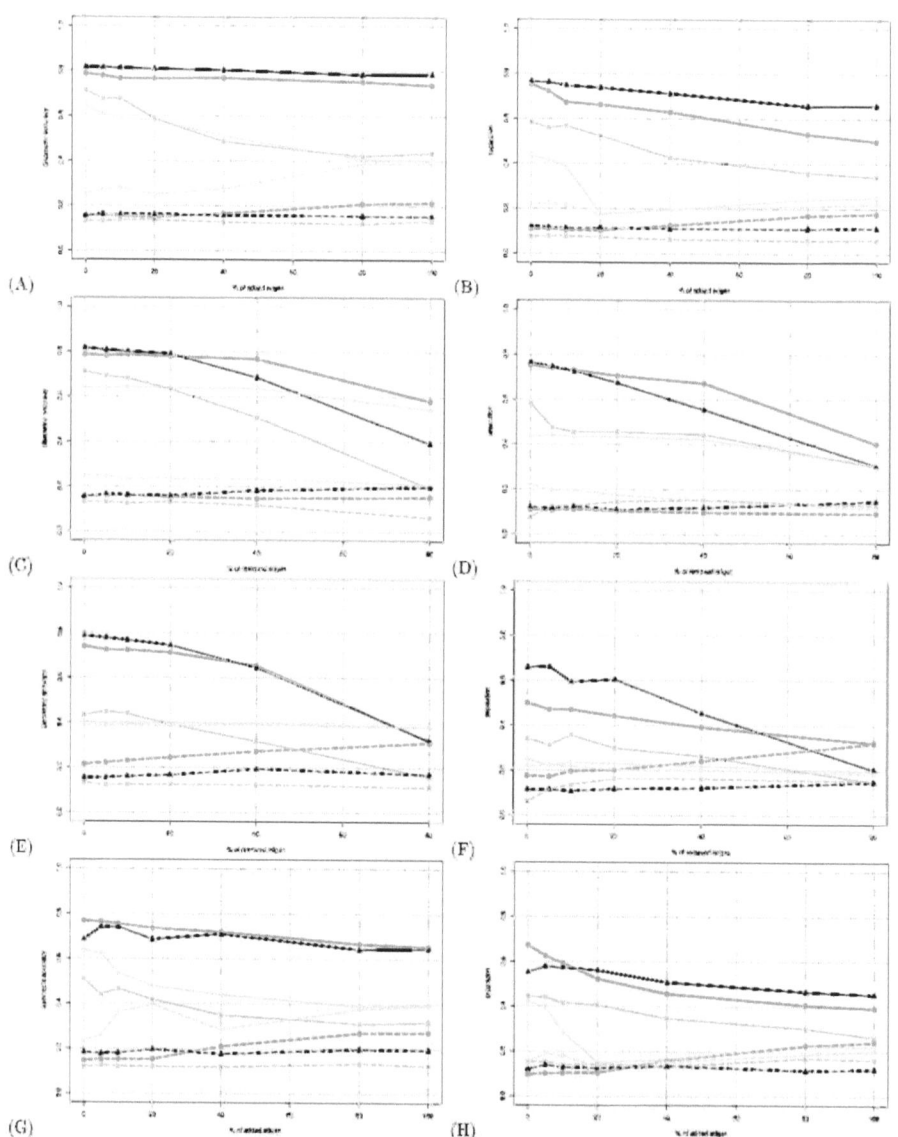

Fig. D.7: Results of the original robustness analysis of clustering algorithm by Brohée *et al.*. Taken from [21]. The quality measures are accuracy (left) and separation (right). (A-B) edge addition to the unmodified graph. (C-D) edge removal from the unmodified graph. (E-F) Edge removal from a graph with 100% randomly added edges. (G-H) Edge addition to a graph with 40% randomly removed edges. The compared approaches are: MCL (blue), RNSC (red), MCODE (orange), SPC (green). The control group is displayed as dotted lines.

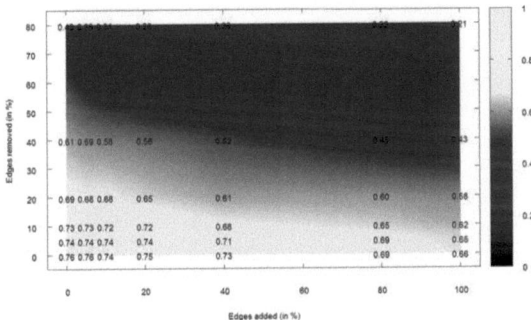

Fig. D.8: Illustration of the results of the robustness evaluation for TransClust. The used quality measure is the separation

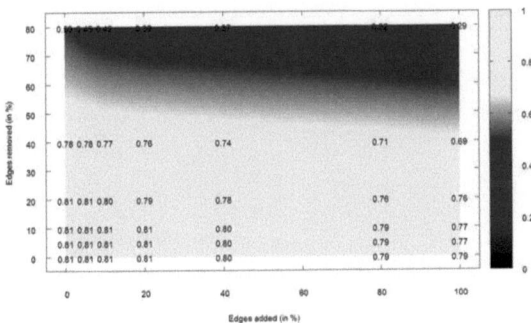

Fig. D.9: Illustration of the results of the robustness evaluation for TransClust. The used quality measure is the accuracy

Die VDM Verlagsservicegesellschaft sucht für wissenschaftliche Verlage abgeschlossene und herausragende

Dissertationen, Habilitationen, Diplomarbeiten, Master Theses, Magisterarbeiten usw.

für die kostenlose Publikation als Fachbuch.

Sie verfügen über eine Arbeit, die hohen inhaltlichen und formalen Ansprüchen genügt, und haben Interesse an einer honorarvergüteten Publikation?

Dann senden Sie bitte erste Informationen über sich und Ihre Arbeit per Email an *info@vdm-vsg.de*.

Sie erhalten kurzfristig unser Feedback!

VDM Verlagsservicegesellschaft mbH
Dudweiler Landstr. 99 Telefon +49 681 3720 174
D - 66123 Saarbrücken Fax   +49 681 3720 1749
www.vdm-vsg.de

Die VDM Verlagsservicegesellschaft mbH vertritt

Printed by Books on Demand GmbH, Norderstedt / Germany